MOLECULAR BEAM EPITAXY

Some Other Pergamon Titles of Interest

PAMLIN: Crystal Growth

KRISHNAN et al: Thermal Expansion of Crystals

WILLIAMS & HALL: Luminescence and the Light Emitting Diode

ROY: Tunnelling and Negative Resistance Phenomena in Semiconductors

TANNER: X-Ray Diffraction Topography

RAY: II-VI Compounds

SHAY & WERNICK: Ternary Chalcopyrite Semiconductors: Growth, Electronic Properties and Applications

SUCHET: Electrical Conduction in Solid Materials (Physicochemical Bases and Possible Applications)

Important Review Journals

Progress in Crystal Growth and Characterization
Editor-in-Chief: B. R. PAMPLIN

Progress in Solid State Chemistry
Editors: G. M. ROSENBLATT & W. L. WORRELL

Free specimen copies available on request

MOLECULAR BEAM EPITAXY

Edited by

Brian R. Pamplin

School of Physics, University of Bath, England

PERGAMON PRESS
OXFORD · NEW YORK · TORONTO · SYDNEY · PARIS · FRANKFURT

U.K.	Pergamon Press Ltd., Headington Hill Hall, Oxford OX3 0BW, England
U.S.A.	Pergamon Press Inc., Maxwell House, Fairview Park, Elmsford, New York 10523, U.S.A.
CANADA	Pergamon of Canada, Suite 104, 150 Consumers Road, Willowdale, Ontario M2J 1P9, Canada
AUSTRALIA	Pergamon Press (Aust.) Pty. Ltd., P.O. Box 544, Potts Point, N.S.W. 2011, Australia
FRANCE	Pergamon Press SARL, 24 rue des Ecoles, 75240 Paris, Cedex 05, France
FEDERAL REPUBLIC OF GERMANY	Pergamon Press GmbH, 6242 Kronberg-Taunus, Pferdstrasse 1, Federal Republic of Germany

Copyright © 1980 Pergamon Press Ltd

All Rights Reserved. No part of this publication may be reproduced, stored in a retrieval system or transmitted in any form or by any means: electronic, electrostatic, magnetic tape, mechanical, photocopying, recording or otherwise, without permission in writing from the copyright holders.

First edition 1980

British Library Cataloguing in Publication Data
Molecular beam epitaxy.
1. Epitaxy 2. Semiconductors
3. Molecular beams
I. Pamplin, Brian Randell
537.6'22   QC611.6.E/   79-41456
ISBN 0-08-025050-5

Published as Volume 2, Number 1/2 of *Progress in Crystal Growth and Characterization,* and supplied to subscribers as part of their subscription.

Printed in Great Britain by A. Wheaton & Co., Ltd., Exeter

CONTENTS

Introduction
B. R. PAMPLIN — 1

Semiconductor superlattices by MBE and their characterization
L. L. CHANG and L. ESAKI — 3

Design considerations for molecular beam epitaxy systems
P. E. LUSCHER and D. M. COLLINS — 15

Nonstoichiometry and carrier concentration control in MBE of compound semiconductors
D. L. SMITH — 33

MBE techniques for IV-VI optoelectronic devices
H. HOLLOWAY and J. N. WALPOLE — 49

Integrated optical devices fabricated by MBE
R. D. BURNHAM and D. R. SCRIFRES — 95

Semiconductor surface and crystal physics studied by MBE
R. Z. BACHRACH — 115

Periodic doping structure in GaAs
G. H. DÖHLER and K. PLOOG — 145

Subject Index — 169

Compound Index — 173

INTRODUCTION

Brian R. Pamplin

Department of Physics, University of Bath, England

Molecular beam epitaxy (MBE) is a new crystal growth technique that has grown out of vacuum evaporation. It is an expensive method but is enjoying a phase of exciting growth because it makes possible the layer by atomic layer growth of compounds and alloys of almost atomically tailored composition. It is superior to vacuum evaporation because that does not produce single crystal layers, let alone tailored compositions. Previous work produced at best only mosaic layers.

MBE is currently being used to produce GaAs devices and $Al_xGa_{1-x}As$ double heterojunction lasers, dielectric quater wave stacks and up to a thousand alternating composition layers for study of the Esaki multijunction negative resistance device. Tunnel diodes, MESFETS (metal semiconductor field effect transistors) and transferred electron devices have been produced. Many III-V compounds and ternary and quaternary alloys have been produced but the technique is not restricted to these compounds. Silicon has been grown - but it is doubtful if this will prove commercially valuable. $Sn_xPb_{1-x}Te$ double heterojunction lasers and detectors are produced and $Sn_{1-x}Pb_xSe$ junctions too. The II-VI compounds have also been grown - ZnSe, ZnTe and $CdSe_{0.7}Te_{0.3}$ on CdS.

A main rival technique is liquid phase epitaxy (LPE) in which crystalline layer junctions and heterojunctions are grown from liquid metal solutions, (Al,Ga)As from Ga solution being the prime example. But MBE layers have proven smoother, more uniform and of larger area. The other rival is vapour phase epitaxy (VPE) which enables III-V compounds, alloys and junctions to be grown epitaxially in a continuous flow system by the use of either volatile halides (usually chlorides) or organometallic compounds to transport the nonvolatile metals like Al, Ga and In. Much finer control of the thickness of successive layers is possible with MBE than with either LPE or VPE and this is where the excitement for the future lies.

Although MBE has been developed from vacuum evaporation and flash evaporation techniques, the use of ultra-high vacua introduces an essentially new feature. Atoms and molecules reach the growth interface not only in very clean conditions but probably in a physically different way. They arrive as simultaneous molecular or atomic beams with nonlocalized properties - like waves rather than particles. This is, I think, not well understood at the present time. From the point of view of classification it may mean that we have here the start of a new main category of crystal growth methods to add to the three existing categories of vapour, liquid and solid growth techniques. (Four if one separates, for convenience, solution growth from melt growth because of the large and diverse list of melt growth techniques.) Perhaps when space processing becomes more common the high vacuum of outer space may be used and techniques based on MBE ideas will be developed for use in an orbital laboratory or a space station. It is also conceivable that three dimensional techniques of tailoring the superlattices of MBE substrates may be found, and the way opened for the construction of inorganic crystals which resemble organic and biological molecules in their complexity.

LPE and VPE are now well documented in the book and review literature (1,2,3,4) although, of course, they are still developing and are a very active area of new research. MBE on the other hand is so new that few review articles (5,6,7) have yet appeared. The first conference devoted solely to MBE was held in Paris in April 1978 and I understand the proceedings are to be published.

The present collection of articles introduces the reader to the use of MBE in the III-V and IV-VI compounds and alloys and indicates the semi-conductor and integrated optics reasons for using the technique. We include articles that show that MBE is capable of wider application. It is the commercial drive for better double heterojunction lasers, detectors, integrated optics and the scientifically interesting possibilities of superlattice structures that has been the driving force behind the rapid progress and has spurred the investment necessary for the research. But just as crystal pulling started as a method for the production of single crystal wires of metals and is now a semiconductor crystal manufacturing technique, so MBE could have a future elsewhere. It may perhaps develop into a technique for growing more complicated metal and inorganic crystals if the need arises. We include, however, just one article on metallic applications of MBE.

REFERENCES

1. Pamplin, B. R. (Ed)(1975) *Crystal Growth*, Pergamon Press (second edition in preparation).
2. Matthews, J. W. (Ed)(1975) *Epitaxial Growth*, Academic Press.
3. International Conferences on Vapour Growth and Epitaxy I to IV, *J. Crystal Growth*, Vols. $\underline{9}$, $\underline{17}$, $\underline{40}$ and to be published.
4. Kaldi, E. and Scheel, H. J. *Crystal Growth and Materials*, European Crystal Growth Conf. I (1976) North Holland (1977).
5. Cho, A. Y. and Arthur, J. R. *Prog. Solid State Chem.* 11., p. 157, Pergamon (1975).
6. Bachrach, R. Z. in (1) to be published.
7. Farrow, R. F. C. in (4) (1977).

SEMICONDUCTOR SUPERLATTICES BY MBE AND THEIR CHARACTERIZATION*

L. L. Chang and L. Esaki

IBM Thomas J. Watson Research Center,
Yorktown Heights, New York 10598, U.S.A.

(Submitted January 1979)

ABSTRACT

The deposition of semiconductor superlattices of $GaAs-Ga_{1-x}Al_xAs$ and $In_{1-x}Ga_xAs-GaSb_{1-y}As_y$ by molecular beam epitaxy is described. Their evaluations by high-energy electron diffraction and X-ray measurement are presented. Properties obtained from transport, optical and magneto experiments are summarized to characterize the electronic subband structure.

1. INTRODUCTION

Superlattices exist in nature. Examples cover a wide range of materials, from the metal alloy of Cu-Au where antiphase boundaries separate regions of different atomic arrangements (1) to the semiconductor of SiC which possesses various polytypical forms (2). The superlattice under consideration in the present context, however, differs from these materials; it refers to a one-dimensional, alternating-layered structure created artificially by periodically depositing two semiconductors.(3) Thus unlike the natural superlattice whose formation is not well understood and is difficult to control, the man-made superlattice can be achieved with predetermined prescriptions in terms of the thickness of the layers, the strength of the periodic potential and the number of periods.

The main objective for making the man-made structure is to investigate a fundamental quantum mechanical phenomenon, the formation of quantized energy states. The period of the superlattice required for this purpose is of the order of 100 Å, which is smaller than the electron mean free path but larger than the lattice spacing. In much the same manner as the lattice potential creates Brillouin zones in the component semiconductors, the superlattice potential subdivides them into minizones, resulting in allowed bands of electron transmission separated by forbidden bands of attenuation. These allowed bands, known as minibands or subbands, are relatively narrow on both energy and wavevector scales with a highly nonlinear dispersion relation. In a superlattice, its configuration determines the structure of the subbands, which in turn dictates the transport and optical properties.

---

*Work sponsored in part under ARO contract.

The original predictions of a negative differential resistance and Bloch oscillations in a superlattice (3) led rapidly to further theoretical considerations on the subject, including electron transport (4,5), electromagnetic waves (6) phonon modes (7) and intrinsic layers (8). But the most challenging task was in the area of experimental implementation to realize such a structure. Early attempts included the fabrication of the GaAs-GaAs$_{1-x}$P$_x$ structure by chemical vapor deposition (9-11) and the GaAs-Ga$_{1-x}$Al$_x$As structure by both molecular beam epitaxy (12) and liquid phase epitaxy(13). The periodic layers thus produced were relatively thick and apparently suffered from nonuniformity and imperfection. Although they did not show the effect of superlattice electronically, they did demonstrate that, at least metallurgically, it was feasible to fabricate a one-dimensional semiconductor structure with a periodic variation in the alloy composition.

The technique of molecular beam epitaxy (MBE) (14,15), which has a number of special and rather unique features, was recognized as the most promising in satisfying the stringent requirements of a superlattice. The relatively low growth temperature minimizes the effect of diffusion. The typically slow growth rate makes possible precise thickness control. The grown surfaces or interfaces are atomically smooth so that extremely thin layers can be produced. Last but not least is its convenience in introducing various beams for compositional modulation. Taking full advantage of these features and adding stable source designs and sophisticated control functions to the MBE system, superlattices of GaAs-Ga$_{1-x}$Al$_x$As with a layer thickness of tens to hundreds of angstroms were achieved (16-18).

Subsequent developments were rapid. Effects of quantum states were observed from transport measurements in single potential wells (19) as well as superlattices (20), and also from optical absorption measurements (21). Additional observations from resonant Raman scattering (22) and magneto-oscillations (23) brought into focus the central feature of the superlattice, the control of subband width and thus its dimensionality. Metallurgical evaluations by X-ray diffraction, meanwhile, established the atomic smoothness of the interfaces (24). Two recent developments helped widening the scope of superlattice activities: The period of the GaAs-AlAs superlattice was reduced to the range of monolayer spacings (25); and other systems such as Si-Si$_{1-x}$Ge$_x$(26) and In$_{1-x}$Ga$_x$As-GaSb$_{1-y}$As$_y$ (27,28) were investigated. The latter system is of particular interest (29). By varying the two alloy compositions independently, the superlattice potential can be changed in strength as well as in character yet the lattice constants can be simultaneously matched.

In this review, we will describe the deposition of superlattices by MBE, and their evaluations by electron diffractions and X-ray measurements. Results of transport and optical experiments will follow to characterize the electronic subband structure. General aspects will be illustrated with GaAs-Ga$_{1-x}$Al$_x$As from which MBE has originated and to which most reports on superlattice properties have been limited. Results of In$_{1-x}$Ga$_x$As-GaSb$_{1-y}$As$_y$ will be described to the extent that they exhibit specific features.

2. SUPERLATTICES BY MBE

Molecular beam epitaxy is basically a vacuum evaporation technique. It refers to the process of epitaxial deposition from molecular beams in an ultrahigh vacuum system. The beams are usually generated thermally in Knudsen-type ovens where quasi-equilibrium is maintained. After expanding in the vacuum space, they condense and grow on the substrate under kinetically controlled conditions. The ovens, each equipped with an individual shutter, are enclosed in a liquid nitrogen shroud. The substrate assembly consists of a holder for mounting the substrate and a heater capable of raising uniformly the substrate to elevated temperatures. Additional accessory equipment for *in situ* cleaning and evaluation may include a mass spectrometer, an Auger analyzer, ion-sputtering facilities and a variety of electron diffraction and microscopy apparatuses.

Although vacuum evaporation has long been used for material preparation, recent interest in applying collimated molecular beams for depositing semiconductor films in ultrahigh vacuum was stimulated by the desorption studies of Ga and As on a GaAs surface (30). As_2 molecules have been found to be reflected from the surface unless Ga atoms, which have a sticking coefficient close to unity, are present. Further investigations have indicated a similar behavior for As_4 molecules (31), and led to the proposal of a kinetic model (32). The As molecules are first adsorbed into a mobile, weakly-bound precursor state and then dissociate upon encountering empty sites provided by Ga. Stoichiometric GaAs can therefore be readily achieved with its growth rate determined by the Ga flux rate as long as an excess of As, be it As_2 or As_4 generated from a GaAs compound or an elemental As, is supplied.

In the growth of $GaAs$-$Ga_{1-x}Al_xAs$ superlattices, four sources are commonly used: Ga, Al, and As for the components and an impurity source such as Sn. Both Al and Sn have apparently a unit sticking coefficient, so that the controls of the alloy composition and the electron concentration are simple and straightforward. The substrate used is typically (100) GaAs, chemically polished and etched. Auger spectroscopy obtained on the etched surface shows contaminations of both oxygen and carbon. The former is easily removed by preheating the substrate to $\sim 500°C$, but the latter can only be eliminated by Ar^+-sputtering. Examinations by electron microscopy show a relatively rough surface, leading to a spotty pattern from glacing-angle observations by high-energy electron diffraction (HEED). The surface begins to smooth out as soon as deposition of GaAs starts, and becomes featureless after a growth of typically ~ 1000 Å. The HEED pattern meanwhile becomes streaked (33) with fraction orders as a result of surface reconstruction (14,15). It is after the completion of the initial smoothing process that the deposition of the super-lattice commences. The deposition conditions are typically: a substrate temperature of 530-580°C, a growth rate of 1-2 Å/sec and a flux ratio of As/Ga or As/(Ga+Al) of 5-10. The surface under these conditions is known as As-stabilized (34), exhibiting a reconstructed c(2x8) structure for both (100) GaAs and $Ga_{1-x}Al_xAs$ with small x, and a disordered (3x1) structure as x is increased to approach AlAs.[17]

The ability of the MBE process to fabricate precise superlattice configurations rests on the stability of the source ovens, the monitor of the beams and the control of the operation. We use a double BN construction for the ovens, containing an inner crucible and an outside tube which is wound with Ta wires for resistive heating. They are completed with double heat-shields made of Ta sheet foils. The beam intensities fluctuate less than ± 2%, corresponding to a temperature variation, for example for the Ga oven, of better than ±1°C (24). For monitor and control, we use a digital computing system in combination with a mass spectrometer (17,18). The computer takes information of flux intensities of the various beams from the mass spectrometer and performs primarily three functions: correction of oven temperatures to maintain constant fluxes; integration of Ga and Al fluxes with respect to time to monitor continuously the thickness of deposition; and operation of the shutters when a predetermined layer thickness is reached. Usually, since the fluxes vary little in operation, only the Al oven is controlled to maintain constant the flux ratio of Al/Ga while both Ga and As ovens are left uncontrolled and their beams exposed to the substrate. The thickness of each layer is automatically adjusted by the opening and closing of the Al shutter alone.

Structural evaluation of the superlattice at the early stage employed the techniques of electron microscopy on cleaved and etched cross sections (18) and nuclear back-scattering from the surface (17,35). While both were satisfactory, resolutions were limited to a layer thickness above ~ 200 Å. The technique of Auger spectroscopy in combination with Ar^+- sputtering successfully profiled structures with a layer thickness of ~ 50 Å, although it failed to provide quantitative information concerning the smoothness and sharpness at the interface(36). The most powerful method of characterization is that of X-ray scattering measurement, including both the large-angle Bragg reflection (37,38) and the small-angle interference (24,37). The former spectrum gives rise to a central peak from the average superlattice as well as to satellite side peaks from the periodic modulation of charge density and lattice constant. The latter spectrum results in a series of interference peaks due to difference in charge density, or rather refractive index, of the periodic layers. Moreover, for a superlattice with only a small number of periods, both primary and secondary interference peaks have been observed. Quantitative analysis of angular positions, intensities and linewidths of these peaks have indicated that the interfacial smoothness and periodic coherency are on the scale of atomic dimensions (24).

Further X-ray measurements have shown that the layer thickness is extremely uniform across the surface of the substrate, and is governed mainly by the geometry of the source ovens. Also, well-defined layers have been observed for a thickness as thin as a few monolayers (39) to a single monolayer (25). The capability of MBE to produce such thin and sharp layers without mixing, as mentioned earlier, is ascribed to its low growth temperature where diffusion effect is expected to be negligible. This has been confirmed by measuring the extremely slow interdiffusion coefficient between Ga and Al from the technique of Auger profiling (40).

Having established the process of making with precision the GaAs-$Ga_{1-x}Al_xAs$ superlattice, its application to the $In_{1-x}Ga_xAs$-$GaSb_{1-y}As_y$ system is rather simple and involves in principle only the replacement of the Al oven with two separate ovens of In and Sb. To control the composition in $GaSb_{1-y}As_y$ is not straightforward, however, for it does not depend linearly on the component fluxes of Sb and As as in the cases of $Ga_{1-x}Al_xAs$ and $In_{1-x}Ga_xAs$ where two group III elements are present (41). Sb, like As, desorbs from the surface unless it reacts with Ga to form GaSb. But it dominates over As when both are present in competing for available sites provided by Ga. The compositional control can best be achieved, therefore, by maintaining the ratio of Sb/Ga below unity. Such dominance of Sb decreases as the substrate temperature is increased in the range of 450-600°C, which is commonly used for the deposition of the $In_{1-x}Ga_xAs$-$GaSb_{1-y}As_y$ superlattice.

The other feature of the $In_{1-x}Ga_xAs$-$GaSb_{1-y}As_y$ system is the additional freedom gained by the use of two alloys, which makes it possible to achieve lattice matching by varying x and y independently (29,42). Both GaAs-AlAs and InAs-GaSb have very close lattice constants with a mismatch of 0.16 and 0.59%, respectively. The mismatch between InAs or GaSb and GaAs, however, is as large as 7-7.5%. The effect of such mismatch on the smoothness of heteroepitaxy has been assessed from *in situ* HEED observations (41). In growing $GaSb_{0.8}As_{0.2}$ on GaAs, for example, the streaked c(2x8) pattern of the latter becomes the streaked c(2x6) pattern which is characteristic of the former only after intervening spotty patterns. This step of intervention, indicative of interruption of planar and smooth growth by nucleation, has been invariably observed for a mismatch in excess of 2.5%. It is important that the $In_{1-x}Ga_xAs$-$GaSb_{1-y}As_y$ superlattice be made with close lattice match, ideally y = 0.92x + 0.08, so that streaked patterns change directly from one to the other to ensure a smooth and abrupt interface.

3. ELECTRONIC PROPERTIES

3.1. GaAs-$Ga_{1-x}Al_xAs$ Superlattice

The semiconductors of GaAs and $Ga_{1-x}Al_xAs$ are technologically important materials which have been investigated extensively. As a superlattice, this combination has received almost exclusive attention not only because of the experimental establishment

of the MBE process for it fabrication but also because it represents a simple and
ideal system from theoretical considerations. GaAs is a direct semiconductor with
a relatively large energy gap. The addition of Al widens the gap but causes little
disturbance otherwise because of similar chemical valence and ion size of Ga and
Al atoms. This widening, occurring mainly in the conduction band (39,43), raises
the conduction bandedge and lowers the valence bandedge on the electron energy
scale, resulting in potential barriers for both electrons and holes. These features
have made it possible to obtain adequately accurate subband structure in most cases
from simple calculations by wavefunction matching, using an effective mass in the
Kronig-Penney model (3,44). Further theoretical calculations have been made (45)
to take into account the effect of band nonparabolicity which is usually small in
this system and the effect of indirect bandgap when the Al composition becomes large
(46). In addition, for structures consisting of only a few monolayers where the
use of bulk-like, effective mass approximation is expected to be invalid, three
dimensional calculations have been performed by both the pseudopotential (47) and
the tight-binding (48) methods.

The initial experimental effort to demonstrate the effect of energy quantization in
a superlattice was naturally centered around the predicted negative resistance
arising from the inflection point in the dispersion relation (3). A relatively
wide subband width was required for this purpose, which was met by using extremely
thin $Ga_{1-x}Al_xAs$ layers (16-18). By reducing the sample area to minimize possible
defects, the negative resistance characteristic was observed at a threshold electric
field expected theoretically. With an improvement in material quality, this threshold has been reduced from $\sim 10^4$V/cm, as originally observed, to the 10^3V/cm range.
However, the magnitude of the negative resistance remained marginal and the
characteristics could not be systematically reproduced.

Subsequent transport measurements were carried out in double-barrier structures(19).
as well as superlattices with a relatively narrow subband width (20). In the former,
essentially discrete quantized states are formed in the GaAs well sandwiched between
two $Ga_{1-x}Al_xAs$ barriers. Resonant transport occurs, giving rise to a current
maximum whenever incident electrons coincide in energy with these states (49,50).
Such resonance was consistently observed at energies in agreement with those calculated
for a variety of structures of different well thicknesses (19,37). Basically
similar, resonant transport was manifested in superlattices with a narrow subband
(20) which becomes misaligned in neighboring wells under a moderate electric field.
Consequently, initial current flow by band conduction gives way to tunneling, (51)
and such current reaches a maximum whenever the ground subband in one well levels
with the second band in an adjacent well. This is expected to lead to an oscillatory
current or conductance as a function of the applied voltage with its period given
by the energy difference between the subbands. Experimentally, such behaviors have
been verified (20).

While transport measurements show primarily the consequences of quantized states
or subbands, these states can be directly probed from optical absorption experiments.
A series of absorption peaks were observed in superlattices of different configurations, corresponding to transitions from subbands in the valence band to those in
the conduction band (21,52). The results have also shown the contributions from
both heavy and light holes and the selection rules for such transitions.

The most informative experiment in relating the subbands and their effect on transport is that of phoconductivity (53). The photocurrent in this case can be measured
using either the photon energy or the applied voltage as the variable. The observed
photo-spectra at a fixed voltage exhibited peaks at energies equal to inter-subband
transitions, similar to the situation in absorption. By setting the photon energy
at one of these transitions and varying the voltage, the current, like that under
d.c. conditions, showed again the negative resistance. Quantitative comparisons

among different superlattices have indicated a reduced conductivity with an increase of the strength of the superlattice potential. Also, measurements as a function of temperature have resulted in expected variations in both the transition energies and the magnitude of the negative resistance.

While these experiments have convincingly demonstrated the formation and control of the subbands, they deal only implicitly with one of the important features of the superlattice: As the superlattice potential increases, the subband width decreases and eventually becomes discrete; and the electrons, which originally have motions associated with three dimensions, are confined in the plane of the layers and become two-dimensional. The density-of-states, as a result, assumes a staircase shape with step rises at these discrete subband energies. From resonant Raman experiments, the observed scattering amplitude, through its dependence on the susceptibility and thus the density-of-states, was greatly enhanced when the photon energies were in resonance with the inter-subband transitions. The calculated spectra in the resonant region for subbands with different degrees of two-to-three-dimensionality fit well the experimental data. Additional Raman measurements have demonstrated the formation of subbands in the spin-orbit valence band, and the operation of Umklapp processes associated with wavevectors which are multiples of the reciprocal period of the superlattice (54).

A more direct and definitive manifestation of the effect of subband dimensionality or anisotropy was obtained from Shubnikov-de Haas oscillations (23,27). The transverse magneto-resistance in this case was measured in the plane of the superlattice, which was of relatively high electron concentration. Since the period of oscillation is a measure of the Fermi cross section perpendicular to the field, it is expected to exhibit different behaviors for subbands of different widths when the field is tilted. Experimental results have borne this out. The period has been found to be independent of the field direction for a wide and nearly three-dimensional subband whose Fermi surface deviates only slightly from being spherical. For a narrow and essentially discrete subband, the dependence becomes a sine function of the tilting angle of the field from the surface normal, which is characteristic of the cylindrical Fermi surface of a two-dimensional electron gas (55). Intermediate variations have also been observed for subbands of intermediate dimensionalities. Quantitative evaluations of these experiments have given further verification to the subband structure: Fermi levels obtained from the periods are in excellent agreement with those calculated on the basis of electron concentrations occupying multiple subbands (23).

3.2. $In_{1-x}Ga_xAs-GaSb_{1-y}As_y$ Superlattice

The semiconductors of InAs and GaAs form a rather interesting combination: Not only are their lattice constants closely matched, as mentioned before, but the conduction bandedge of the former lies even below the valence bandedge of the latter on the electron energy scale (29,56,57). This energy difference, referred to as bandedge separation and defined as negative in this case, can be made close to zero near compositions of x, $y \cong 0.3$ and positive thereafter, when these semiconductors are alloyed to become $In_{1-x}Ga_xAs$ and $GaSb_{1-y}As_y$. Recognition of this rather unique property has led us to initiate the MBE depositions of these materials (41), uncover quite unusual rectifying-to-ohmic transport characteristics for heterojunctions (42), and propose a three-terminal tunneling device structure (58).

As a superlattice, the $In_{1-x}Ga_xAs-GaSb_{1-y}As_y$ system can be distinguished in general from the $GaAs-Ga_{1-x}Al_xAs$ system in that the former has the conduction and valence bandedges of one alloy lying above the corresponding edges of the other. It follows that $In_{1-x}Ga_xAs$ and $GaSb_{1-y}As_y$ layers serve as potential wells for electrons and holes, respectively, in contrast to the previous situation where GaAs layers serve as wells for both carriers. This feature of spatial separation is expected to have significant influence on optical absorption and photoconductivity properties. More importantly, the mixing or interaction of the conduction and valence bands makes it invalid to calculate the subband structure by simple plane-waves. Appropriate Bloch wavefunctions in the framework of the kp method (59), instead, have been used for this purpose (29). The results show a number of interesting features, all of course arising from the present, unusual bandedge relationship. The inflection point in the dispersion curve of the subband, in general, moves toward the zone center, as compared to the $GaAs-Ga_{1-x}Al_xAs$ system, which is desirable in terms of reaching the negative mass region. The energy gap of the superlattice, defined as the energy difference between the fundamental subbands of electrons and heavy holes, can be controlled over a wide range. Unlike the situation in $GaAs-Ga_{1-x}Al_xAs$ where it is limited to the region between the energy gaps of the two components, the superlattice gap in the present system can be made smaller than either of them. For negative bandedge separation as in InAs-GaSb, indeed, it may be zero and the subbands of electrons and holes may cross each other, resulting in a semimetallic state. Alternative calculations have also been carried out, including the method of linear combination of atomic orbitals (60) and a three-dimensional computation mainly for thin monolayer structures (61). The results are all in agreement with one another for configurations in the range of our interest.

The superlattices of InAs-GaAs with relatively thick layers were first investigated experimentally, and their essentially two-dimensional subbands were established from magneto-oscillations (28) similarly as before. The favorable electron mobility in the present system has, in addition, unabled us to observe two sets of oscillations associated with both the fundamental and the second subbands. The upper subband, in other words, has made its presence felt not merely through its occupation by electrons alone, as in the case of $GaAs-Ga_{1-x}Al_xAs$. The effect of band nonparabolicity, not important in the previous system, becomes significant here and results in a modified density-of-states and an enhanced electron mass, as observed experimentally from the temperature dependence of the amplitude of oscillations (28). Such enhancement, as a function of energy from the conduction bandedge of InAs, is identical with that of bulk films. This is consistent with, and indeed gives further support to, the two-dimensional nature of the electrons which are essentially confined in the InAs layers.

The measurements of optical absorptions were carried out to determine subband energy positions and to ascertain theoretical calculations.(62) The observed intensities were rather weak, reflecting the spatial separation of carriers as mentioned earlier. Well-defined absorption edges, however, have been obtained from a large number of superlattices including both the pure compounds, InAs-GaSb, and the alloys, $In_{1-x}Ga_xAs-GaSb_{1-y}As_y$. In the former structure, a detailed comparison between these results and the calculated superlattice gaps has led to the determination of, perhaps, the most important parameter of the present system, that of the bandedge separation defined previously. It is indeed negative and falls in the vicinity of -0.15eV. This value is close to that deduced from electron affinities (56), although both positive and negative values have been predicted theoretically (57,63). Using this number and the expected variations of bandedges with alloy compositions in $In_{1-x}Ga_xAs-GaSb_{1-y}As_y$ (29,42), the energy gaps of superlattices with different bandedge separations have been calculated and found to agree well with the absorption edges obtained experimentally. In all cases, they are smaller than the energy gaps of the component materials, a unique characteristic of this type of superlattice as described before.

At the present time, a number of experiments are in progress. Refined absorption measurements exhibit structured spectra which appear to correlate with additional, heavy-hole subbands. Under magnetic fields, the spectra show apparently interesting, although not completely understood, oscillatory characteristics. Transport measurements with different layer thicknesses, on the other hand, seem to provide evidence to the predicted semimetallic transition. While all these experiments are clearly preliminary, they do point to the ever expanding and stimulating nature of the superlattice.

ACKNOWLEDGMENT

We acknowledge valuable contributions from many of our colleagues and visiting scientists, with whom we have been privileged to work together in the pursuance of the superlattice. In particular, we are grateful to R. Tsu, R. Ludeke, G. A. Sai-Halasz, and C. A. Chang; and to H. Sakaki of the University of Tokyo.

REFERENCES

1. Von C. H. Johansson and J. O. Linde, *Phys.* **25**, 1 (1936).

2. See, for example, A. R. Verma, *Crystal Growth and Dislocations*, Butterworths, London, 1953.

3. L. Esaki and R. Tsu, *IBM J. Res. Develop.* **14**, 61 (1970).

4. P. A. Lebwohl and R. Tsu, *J. Appl. Phys.* **41**, 2664 (1970).

5. P. J. Price, *IBM J. Res. Develop.* **17**, 39 (1973).

6. R. F. Kazarinov and R. A. Suris, *Soviet Phys. Semicond.* **6**, 120 (1972) *(Fiz. Tekh. Poluprov.* **6**, 148, 1972).

7. R. Tsu and S. Jha, *Appl. Phys. Lett.* **20**, 16 (1972).

8. G. H. Dohler, *Phys. Stat. Sol.* (b), **52**, 533 (1972).

9. A. E. Blakeslee and C. F. Alliotta, *IBM J. Res. Develop.* **14**, 686 (1970).

10. L. Esaki, L. L. Chang, and R. Tsu, in *Proc. 12th Int. Conf. Low Temp. Phys.* Kyoto, 1970, p. 551.

11. Zh. I. Alferov, Yu. V. Zhilyaev, and Yu. V. Shmartsev, *Soviet Phys. Semicond.* **5**, 174 (1971). *(Fiz. Tekh. Poluprov.* **5**, 196, 1971).

12. A. Y. Cho, *Appl. Phys. Lett.* **19**, 467 (1971).

13. J. W. Woodall, *J. Crystal Growth* **12**, 32 (1972).

14. L. L. Chang and L. Ludeke, in *Epitaxial Growth*, ed. by J. W. Matthews (Academic, New York 1975), part A, p. 37.

15. A. Y. Cho and J. R. Arthur, in *Progress in Solid State Chemistry* ed. by J. O. McCaldin and G. A. Somorjai (Pergamon, New York 1975) Vol. 10, p. 157.

16. L. Esaki, L. L. Chang, W. E. Howard, R. Ludeke, and V. L. Rideout, in *Proc. 11th Int. Conf. Phys. Semicond.*, Warsaw 1972 (PWN-Polish Scientific, Warsaw 1972) p. 431.

17. L. L. Chang, L. Esaki, W. E. Howard and R. Ludeke, *J. Vac. Sci. Technol.* **10**, 11 (1973).

18. L. L. Chang, L. Esaki, W. E. Howard, R. Ludeke, and G. Schul, *J. Vac. Sci. Technol.* 10, 655 (1973).

19. L. L. Chang, L. Esaki, and R. Tsu, *Appl. Phys. Lett.* 24, 593 (1974).

20. L. Esaki and L. L. Chang, *Phys. Rev. Lett.* 33, 495 (1974).

21. R. Dingle, W. Wiegmann, and C. H. Henry, *Phys. Rev. Lett.* 33, 827 (1974).

22. P. Manuel, G. A. Sai-Halasz, L. L. Chang, C.-A. Chang, and L. Esaki, *Phys. Rev. Lett.* 37, 1701 (1976).

23. L. L. Chang, H. Sakaki, C.-A. Chang, and L. Esaki, *Phys. Rev. Lett.* 38, 1489 (1977).

24. L. L. Chang, A. Segmüller and L. Esaki, *Appl. Phys. Lett.* 28, 39 (1976).

25. A. C. Gossard, P. M. Petroff, W. Wiegmann, R. Dingle, and A. Savage, *Appl. Phys. Lett.* 29, 323 (1976).

26. E. Kasper, H. J. Herzog and H. Kibbel, *Appl. Phys.* 8, 199 (1975).

27. L. L. Chang, *Surf. Sci.* 73, 226 (1978).

28. H. Sakaki, L. L. Chang, G. A. Sai-Halasz, C.-A. Chang, and L. Esaki, *Solid State Commun.* 26, 589 (1978).

29. G. A. Sai-Halasz, R. Tsu, and L. Esaki, *Appl. Phys. Lett.* 30, 651 (1977).

30. J. R. Arthur, *J. Appl. Phys.* 39, 4032 (1968).

31. C. T. Foxon and B. A. Joyce, *Surf. Sci.* 50, 434 (1975).

32. J. R. Arthur, *Surf Sci.* 43, 449 (1974).

33. D. B. Dove, R. Ludeke, and L. L. Chang, *J. Appl. Phys.* 44, 1897 (1973).

34. A. Y. Cho, *J. Appl. Phys.* 42, 2074 (1971).

35. J. W. Mayer, J. F. Ziegler, L. L. Chang, R. Tsu, and L. Esaki, *J. Appl. Phys.* 44, 2322 (1973).

36. R. Ludeke, L. Esaki, and L. L. Chang, *Appl. Phys. Lett.* 24, 417 (1974).

37. L. L. Chang, L. Esaki, A. Segmüller and R. Tsu, in *Proc 12th Int. Conf. Phys. Semicond.* ed. by M. H. Pilkuhn, Stuttgart, 1974 (B. G. Teubuer, Stuttgart, 1974) p. 688.

38. A. Segmüller, P. Krishna, and L. Esaki, *J. Appl. Crystallogr.* 10, 1 (1977).

39. L. Esaki and L. L. Chang, *Thin Solid Films* 36, 285 (1976).

40. L. L. Chang and A. Koma, *Appl. Phys. Lett.* 29, 138 (1976).

41. C.-A. Chang, R. Ludeke, L. L. Chang, and L. Esaki, *Appl. Phys. Lett.* 31, 759 (1977).

42. H. Sakaki, L. L. Chang, R. Ludeke, C.-A. Chang, G. A. Sai-Halasz, and L. Esaki, *Appl. Phys. Lett.* 31, 211 (1977).

43. R. Dingle, in *Adv. Solid State Phys.*, ed. by H. J. Queisser (Vieweg, Braunschweig 1975), Vol. 15, p. 21.

44. See, for example, R. A. Smith, *Wave Mechanics of Crystalline Solids* (Wiley, New York, 1961), p. 138.

45. D. Mukherji and B. R. Nag, *Phys. Rev. B*, 12, 4338 (1975).

46. H. C. Casey and M. B. Panish, *J. Appl. Phys.* 40, 4910 (1969).

47. E. Caruthers and P. J. Lin-Chung, *Phys. Rev. Lett.* 38, 1543 (1977).

48. J. N. Schulman and T. C. McGill, *Phys. Rev. Lett.* 39 1680 (1977).

49. E. O. Kane, in *Tunneling Phenomenon in Solids*, ed. by E. Burstein and S. Lundqvist (Plenum, New York, 1969), Chap. I.

50. R. Tsu and L. Esaki, *Appl. Phys. Lett.* 22, 562 (1973).

51. R. Tsu and G. H. Dohler, *Phys. Rev. B* 12, 680 (1975).

52. R. Dingle, A. C. Gossard and W. Wiegmann, *Phys. Rev. Lett.* 34, 1327 (1975).

53. R. Tsu, L. L. Chang, G. A. Sai-Halasz, and L. Esaki, *Phys. Rev. Lett.* 34, 1509 (1975).

54. G. A. Sai-Halasz, A. Pinczuk, P. Y. Yu, and L. Esaki, *Solid State Commun.* 25, 381 (1978).

55. F. F. Fang and P. J. Stiles, *Phys. Rev.* 174, 823 (1968).

56. G. W. Gobelli and F. G. Allen, *Phys. Rev.* 137, A245 (1965).

57. W. A. Harrison, *J. Vac. Sci. Technol.* 14, 1016 (1977).

58. L. L. Chang and L. Esaki, *Appl. Phys. Lett.* 31, 687 (1977).

59. E. O. Kane, *J. Phys. Chem. Solids* 1, 249 (1957).

60. G. A. Sai-Halasz, L. Esaki, and W. A. Harrison, *Phys. Rev.*, in press.

61. R. A. Nucho and A. Madhukar, presented at 5th Conf. Phys. Compound. Semicond. Interfaces, Los Angeles, 1978 (*J. Vac. Sci. Technol.*, in press).

62. G. A. Sai-Halasz, L. L. Chang, J. W. Welter, C.-A. Chang, and L. Esaki, *Solid State Commun.*, in press.

63. W. R. Frensley and H. Kroemer, *Phys. Rev. B* 16, 2642 (1977).

THE AUTHORS

L. L. Chang

L. Esaki

Dr. L. L. Chang received his B.S. degree (1957) from the National University of Taiwan, his M.S. degree (1961) from the University of South Carolina, and his Ph.D. from Stanford University (1963). Between 1963 and 1968 he was at the IBM Thomas J. Watson Research Center, Yorktown Heights, New York; he returned in 1969, and is currently engaged in research there. Between 1968 and 1969 he held an Associate Professorship at M.I.T., Cambridge, Massachusetts.

Dr. Chang is a member of the American Physical Society, the American Vacuum Society, the IEEE, and Sigma Xi.

Dr. Leo Esaki received his B.S. and Ph.D. degrees in Physics from the University of Tokyo, in 1947 and 1959, respectively. He is an IBM Fellow and has been engaged in research at the IBM Thomas J. Watson Research Center, Yorktown Heights, New York since 1960. Prior to joining IBM, he worked at the Sony Corp., where his research on heavily-doped Ge and Si resulted in the discovery of the tunnel diode. His major field is semiconductor physics. Dr. Esaki's current interest includes a man-made semiconductor superlattice in search of a predicted quantum mechanical effect.

Dr. Esaki was awarded the Nobel Prize in Physics (1973) in recognition of his pioneering work on tunneling in solids and discovery of the tunnel diode. Other awards include the Nishina Memorial Award, the Asahi Press Award, an achievement award from Tokyo Chapter of the U.S. Armed Forces Communications and Electronics Association, the Toyo Rayon Foundation Award for the Promotion of Science and Technology, the Morris N. Liebmann Memorial Prize, the Stuart Ballantine Medal and the Japan Academy Award in 1965. Dr. Esaki holds honorary degress from Doshisha School, Japan, and the Universidad Politecnica de Madrid, Spain. From 1971 to 1975 he served as Councillor-at-Large of the American Physical Society, and as a Director of the American Vacuum Society from 1973 to 1975. Dr. Esaki is a Director of IBM-Japan, Ltd., and a Director of the Yamada Science Foundation. He serves on numerous international scientific advisory boards and committees. Currently he is a Guest Editorial writer for the Yomiuri Press. He was chosen for the Order of Culture by the Japanese Government in 1974. Dr. Esaki was elected a Fellow of the American Academy of Arts and Sciences in May 1974, a member of the Japan Academy on 12 November 1975, a Foreign Associate of the National Academy of Sciences (USA) on 27 April 1976, and a Foreign Associate of the National Academy of Engineering (USA) on 1 April 1977. On 31 May 1978, Dr. Esaki was elected a Corresponding Member of the Academia Nacional De Ingenieria of Mexico.

Dr. Esaki is a Fellow of the American Physical Society and of the IEEE, and he is a Member of the Physical Society of Japan, and of the Institute of Electrical Communication Engineers of Japan.

DESIGN CONSIDERATIONS FOR MOLECULAR BEAM EPITAXY SYSTEMS

P. E. Luscher and D. M. Collins

Varian Associates Inc., Palo Alto, CA 94303, U.S.A.

(Submitted December 1978)

ABSTRACT

Important factors which must be considered in the design of a molecular beam epitaxy (MBE) system are addressed. Vacuum system requirements are discussed with regard to film purity levels. The important factors involved in the design of molecular beam sources (flux uniformity, crucible volume, temperature stability and reproducibility, furnace and crucible materials, heat shielding, source baffling, and shutters) and substrate holders (temperature stability, reproducibility, and uniformity, and the mounting of substrates with liquid metal films) are investigated.

The use of *in situ* analytical equipment (e.g. reflection electron diffraction, quadrupole mass spectroscopy, and Auger electron spectroscopy) is discussed and the importance of these techniques in both experimental and production MBE systems is evaluated.

1. INTRODUCTION

Molecular beam epitaxy (MBE) is rapidly establishing itself as a technology of great potential importance for the fabrication of microwave and optoelectronic devices(1-4). Much of the recent progress in growing device quality epitaxial semiconductor films by MBE was made possible by the increasing availability of commercial ultrahigh vacuum systems and surface analysis equipment. Using these building blocks, scientists developed systems suited to growth studies of a wide range of semiconductor materials. As a result of these scientific studies, MBE technology has matured over the last 10 years to the extent that commercial MBE systems are now available in the market place. The purpose of this article, is to describe the individual components of an MBE system and the manner in which they are combined to achieve the ultimate goal of growing high quality epitaxial layers with precisely controlled thicknesses and doping profiles.

2. THE MBE PROCESS

The requirements a MBE system must meet are best understood by defining the MBE process and giving a brief description of the growth kinetics which make epitaxial growth by MBE feasible for a wide range of materials.

The MBE process achieves epitaxial growth in an ultra-high vacuum (UHV) environment through the chemical reaction of multiple molecular beams of differing flux densities with a heated single crystal substrate. This process is illustrated schematically in Fig. 1 which shows the essential components necessary for MBE growth of doped (AlGa)As. The illustration borders represent the confines of the UHV system. Each furnace contains a crucible which in turn contains one of the constituent chemical elements or compounds of the desired film. The temperature of each furnace is chosen so that the vapor pressures of the materials are sufficiently high for generation of thermal energy molecular "beams" by free evaporation. The furnaces are arranged so that the central portion of the beam flux distribution from each furnace intersects the substrate. By choosing appropriate furnace and substrate temperatures, epitaxial films of the desired chemical co-position can be obtained. Additional control over the growth process is achieved by individual shutters interposed between each furnace and the substrate. Operation of these shutters permits abrupt cessation or initiation of any given beam flux to the substrate.

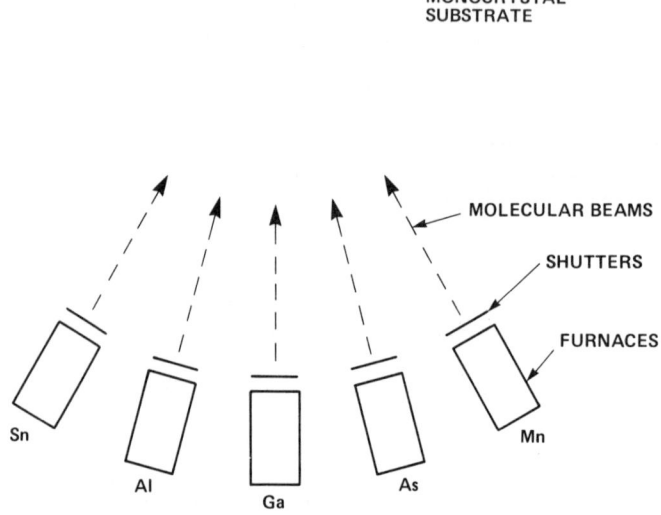

Fig. 1. Schematic illustration of a system configuration for growth of doped $Al_xGa_{1-x}As$ by MBE. Tin functions as a donor impurity (n-type dopant) and Mn functions as an acceptor impurity (p-type dopant).

One of the distinguishing characteristics of MBE is the low growth rate: approximately 1 μm h^{-1} or, equivalently, 1 monolayer $^{-1}$ The molecular beam flux at the substrate can therefore be readily modulated in monolayer quantities with shutter operation times below 1s.

This incredible precision in beam flux control must combine with negligible bulk diffusion and atomically smooth surfaces if structures with composition profiles precise to ~ 5 Å are to be realized. That all of the necessary criteria for atomic level control over the geometrical and chemical parameters in the growth direction can be met is most beautifully demonstrated by the growth of superlattice materials (5-7) which are discussed in detail in pp.3-14 of this issue.

In addition to satisfying geometrical parameters the films must exhibit the desired electrical transport and/or optical properties. These properties depend primarily upon film purity; type, lattice site and concentration of dopant; concentration and structure of defects; and stoichometry of the host lattice material. These properties in turn depend to varying degrees upon the growth kinetics.

In the case of films of Group IV elements, the kinetics of MBE growth appear straightforward. However, for compound semiconductor materials in which neither molecular nor congruent evaporation of the compound occurs* it is not *a priori* obvious that a means of growing stoichiometric material exists. Molecular evaporation has been observed in IV-VI compounds (e.g., PbTe (2,8) and congruent evaporation in II-VI compounds (e.g., CdTe(2) and ZnTe(8). In III-V compounds, molecular evaporation of the compounds has not been observed and the beam flux at temperatures where congruent evaporation occurs is too low to be of practical use(2). Flux matching from separate elemental sources has practical limitations and cannot be considered as a solution. In fact, since growth is ultimately determined by complex kinetic reaction processes at the substrate surface, even molecular beams generated by congruent evaporation need not result in stoichiometric epitaxial films. However, fundamental studies of adsorption-desorption kinetics(9) have demonstrated that a solution to the stoichoimetry problem in the III-V compounds lies in the surface chemical dependence of the sticking coefficient of the Group V elements. At temperatures at which epitaxial growth occurs, only that amount of the Group V element is adsorbed which satisfies the available Group III bonding orbitals at the surface. The growth rate is therefore determined by the arrival rate of the Group III elements while the condition of stoichiometry is satisfied simply by growing in an excess flux of the Group V elements. While this is a simplified description of what is in reality a complex process, the validity of this model has made possible growth of high quality III-V epitaxial layers on III-V single crystal substrates by MBE.

An inherent and advantageous property of MBE which is lacking in liquid phase epitaxy (LPE) and vapor phase epitaxy (VPE) is that the vacuum environment is ideal for *in situ* analytical instrumentation. While this will be discussed in more detail below, it should be mentioned here that the incorporation of molecular beam and surface analytical equipment in a single UHV system was responsible for much of the early progress in defining and understanding MBE growth mechanisms.

3. THE MBE SYSTEM

While Fig. 1 is sufficient for discussion of the MBE process, its actual implementation is considerably more complex than Fig. 1 might imply. Figure 2 is a photograph of an MBE system which comprises the following components:

- An UHV system, including a specimen exchange vacuum load-lock
- A heated substrate holder
- Multiple furnaces
- Furnace shutters
- Furnace baffles
- Epitaxy control instrumentation
- Components for *in situ* substrate cleaning
- Surface analytical instrumentation

---

* Molecular evaporation as used in this text refers to vaporization of the binary compound, e.g. PbTe (solid) → PbTe (gas). Congruent evaporation refers to vaporization in which the elemental concentration of the molecular beam equals that of the solid, e.g. CdTe (solid) → Cd (gas) + 1/2 Te_2 (gas). For a complete discussion and review of vaporization of the III-V, II-VI and IV-VI binary semiconductors, see Ref. (2).

An exploded view of the work chamber of the MBE system pictured in Fig. 2 is shown in Fig. 3. The right-hand side of the work chamber contains surface analytical instruments used sequentially with the growth process. The left-hand side contains the components used directly in the growth process as well as two analytical instruments reflection electron diffraction (RED or HEED) and a quadrupole mass spectrometer (QMS), which may be used simultaneously with the growth process. The main shutter effectively splits the chamber in half to prevent deposition on the analytical components which are not used during growth. The growth components consist of a heated substrate holder, multiple evaporation furnaces, baffles, and furnace shutters.

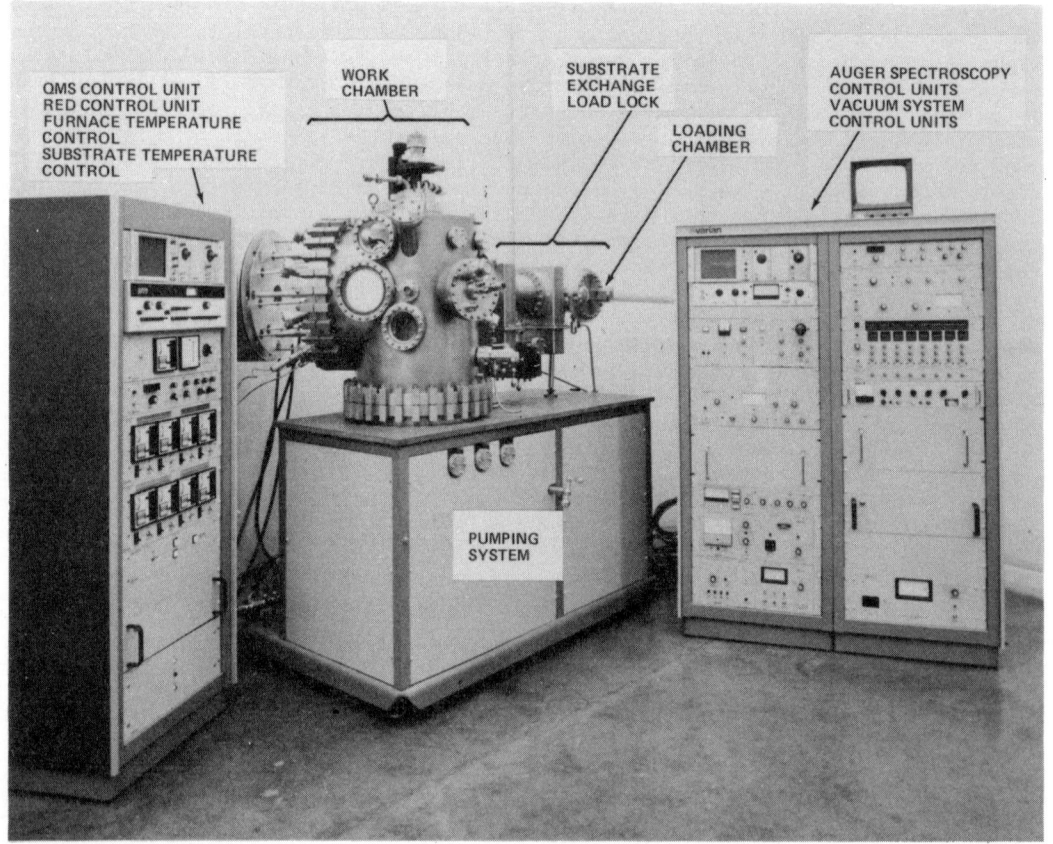

Fig. 2. The molecular beam epitaxy system pictured in this photograph contains all of the components needed for growth and in situ analysis of epitaxial semi-conductor films. (Courtesy of Varian Associates, Inc.)

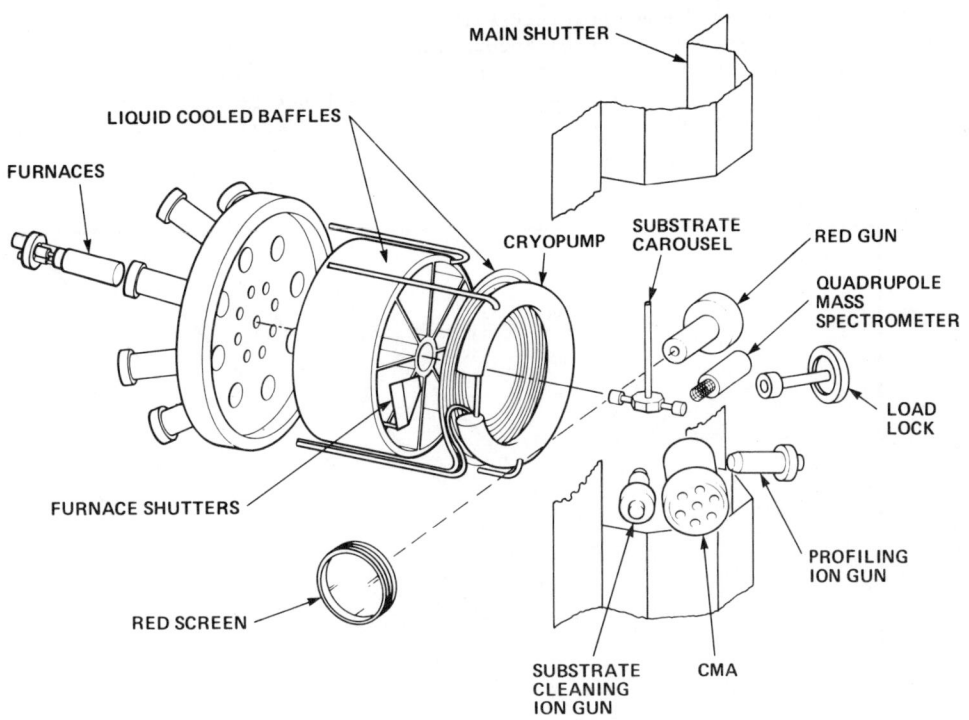

Fig. 3. This exploded illustration describes the components and their configuration in the work chamber of the MBE system shown in Fig. 2. (Courtesy of Varian Associates, Inc.)

3.1 The UHV system

The need for UHV conditions arises from the low growth rates coupled with the obvious requirements of negligible unintentional film impurity levels. The concentration of unintentional contaminants incorporated into a growing film depends on the sticking coefficients (ξ_i, on the substrate at growth temperature), the partial pressures (P_i (Torr)) and the molecular weights (M_i (AMU)), of the background gas species i; as well as the rate of growth (r_g (cm/s)) of the epitaxial film. The total contamination incorporated in the growing film, at concentrations $< 10^{21}$ cm^{-3}, may be written as

$$C = \Sigma_i\, C_i = \Sigma_i\, \xi_i\, \frac{d^2N_i/dtds}{r_g} \qquad (1)$$

where C_i is the impurity concentration of gas molecules of specie i per cm^3 and $d^2N_i/dtds$ is the number of molecules of gas specie i which strike the growth surface per cm^2 per s. From kinetic gas theory $d^2N_i/dtds$ may be written

$$\frac{d^2N_i}{dtds} = 3.5 \times 10^{22}\, \frac{P_i}{(M_i T)^{1/2}} \qquad (2)$$

where T is the background gas temperature (K)(10). In Fig. 4 we present the concentration of CO molecules (M_{CO}=28 AMU) incorporated into an epitaxial film as a function of CO partial pressure (P_{CO}(Torr)) and sticking coefficient (ξ_{CO}) assuming a growth rate of 1 μm h^{-1}. It is clear from Fig. 4 that the product $P_{CO}\,\xi_{CO}$ must be $\lesssim 7.3 \times 10^{-16}$ torr to insure a CO molecular impurity concentration less than 10^{13} cm^{-3}.

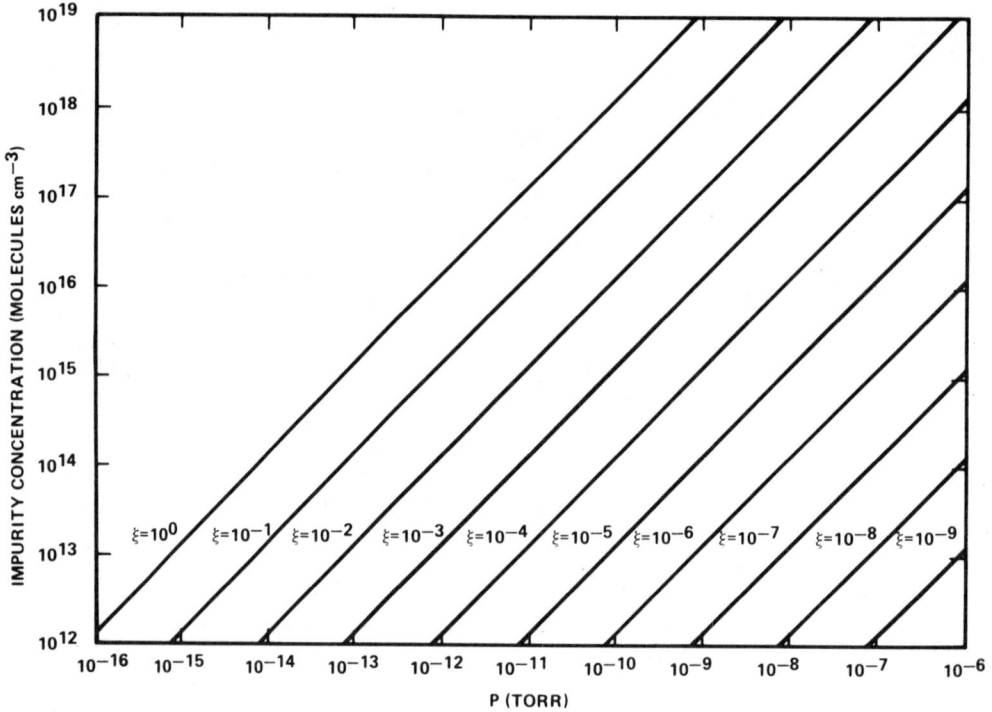

Fig. 4. The calculated impurity concentration of CO molecules as a function of CO partial pressure for a growth rate of 1 μm h^{-1}. The parameter ξ is the sticking probability of CO on the substrate at growth temperature.

Base total pressures routinely achieved in UHV systems are in the low 10^{-10} to mid 10^{-11} torr range, and this background pressure increases somewhat during operation due to the increased heat load from the furnaces and substrate. It is immediately apparent that the present success of MBE technology stems as much or more from the small sticking coefficients of the residual gas species on the heated substrate as from the residual gas pressure. MBE UHV system design is focused on minimizing total system background pressure, eliminating gas species with high sticking coefficients, and eliminating those gas species which are electically active if incorportated into the epitaxial film.

It should be pointed out that the purity of the source material may be extremely important. These materials are seldom more than 99.9999% pure and hence it is vital that the vapor pressures of the impurities are either much greater (so they may be rapidly outgassed) or much lower than that of the source material itself.

Design Considerations for MBE Systems

If this is not the case and the sticking coefficients of the impurities on the film are significant, no amount of pumping speed will be of help.

To obtain UHV conditions, MBE systems are constructed of low vapor pressure materials which can be baked at 200°C and greater for extended periods of time (8-24 h) to accelerate outgassing. Pumping of the system is achieved with a combination of different pumping mechanisms. Rough pumping is achieved by LN_2 cooled sorption pumps and carbon vane pumps. For UHV, titanium sublimation, LN_2 cryosorption, and sputter ion pumping are generally used, although LN_2 trapped diffusion pumps (3) and closed-loop helium cryopumps (11) have been employed. In Fig. 2 the pumping system is housed in the rectangular cabinet beneath the work chamber.

The procedure for attaining UHV conditions is too time consuming to permit exposure of the vacuum system to atmospheric pressure simply for substrate exchange. In addition, exposure to atmospheric pressure can cause oxidation or other contamination of the furnace charges and hence require careful outgassing of the furnaces and charges prior to subsequent growths. Substrate exchange laod-locks have therefore become an integral part of MBE vacuum systems. They permit operation over periods of weeks, or even months, during which the vacuum remains unbroken. Futhermore, the capability which this provides for operating furnaces continuously at reduced temperatures between growths may improve crucible life for some materials as well as ultimate film purity levels.

For a load-lock system to be most effective, the transfer process should introduce a minimum of contamination into the work chamber and maximize the time during which the system is under actual growth operation. Such a system is shown in Fig. 5. Since typical growth times for MBE films are greater than 30 min, load-lock systems which provide the facility to introduce and evacuate one substrate in the loading chamber while growth is occurring on another substrate in the work chamber are most desirable. At the end of the growth the two substrates can be quickly interchanged and the delay reduce to just a few minutes. It is important that the increase in the background pressure in the work chamber during substrate transfer be small enough so that return of the work chamber background pressure to pretransfer levels requires only a few minutes. Typically an increase in work chamber background pressure of 10^{-8} to 10^{-7} torr can be tolerated furing the transfer. The longer the loading chamber is under vacuum, the smaller will be the pressure rise during exchange. Once the substrate and its mounting block are heated to the growth temperature, they are thoroughly degassed and contribute minimally to the residue gas spectrum during growth. Elements of the transfer mechanism and the loading chamber which are routinely exposed to atmospheric pressure are valved off from the work chamber during growth.

3.2 The heated substrate holder

To insure uniform and reproducible film properties it is important that the substrate be mounted in such a manner that the temperature across the substrate is uniform and reproducible within $\pm$ 10°C. It is also important that the mounting of the substrate does not subject it to undue stress or to impurities which may incorporate into the growing film(12). For III-V compound semiconductors the substrate wafer is generally mounted with indium to a uniformly heated block. Uniform temperature across the mounting block is generally accomplished by making the block with thick cross section from a refractory material with a high thermal conductivity such as molybdenum(1). It has also been accomplished using a uniformly distributed heater (e.g. a deposited Ta film) on an insulating block (e.g. quartz, beryllia or sapphire)(3,12). Reduction of strong direct current and particularly alternating current fields generated by the heater element is important to minimize distortion of the RED pattern and Auger spectrum. The indium, which is liquid at the growth temperature, holds the wafer to the heated mounting block by surface tension and provides excellent heat transfer

Fig. 5. A close-up view of the MBE substrate exchange vacuum load-lock for the MBE system shown in Fig. 2. (Courtesy of Varian Associates, Inc.) Operation of the load-lock proceeds as follows:
(1) The loading chamber and transfer drive are backfilled with dry N_2 to atmospheric pressure.
(2) A substrate heater block with the substrate mounted in position is placed on the end of the transfer drive in the loading chamber. (3) The loading chamber and transfer drive are rapidly rough pumped to < 10 μm Hg. (4) The swing valve on the right is opened to permit the load-lock ion pump to pump on the load chamber and transfer drive. (5) After the load-lock ion pump pressure drops below $\sim 5 \times 10^{-6}$ torr, the second swing valve may be opened and the transfer may be made.

from the heater block. While gallium has been used in place of indium, it has been found to attack and dissolve in the molybdenum block(1). In the growth of IV-VI compound semiconductors, tin has been used for mounting the substrate(13). The vapor pressures of all three of these liquids are quite low at the growth temperatures at which they are used (typically $\leq$ 600°C) and, when properly applied, have not been observed to contaminate the growing films. Clips have been used to attach the substrate to the mounting block, but they tend to induce dislocations in the vicinity of their contact with the crystal and substrate temperature measurement and uniformity are significantly more difficult to achieve(1). The higher substrate temperatures required for Si MBE (750-1100°C) preclude the use of a liquid metal bonding agent as is done with the II-VI, IV-VI and III-V compound semiconductors. However, direct radiation heating from a uniform high temperature radiator immediately behind the substrate has been successfully employed (14,15) as has the use of specially shaped Si wafers held in Ta clips and heated by passing a current through the wafer itself (16,17). It is much more difficult to achieve uniform heating and eliminate strain induced by the Ta clips when heating by the latter method.

Measurement of the substrate temperature may be accomplished in several ways. Most common is the use of a thermocouple either placed in a well in the back of the substrate, soldered to a dummy substrate, or in contact with the substrate mounting block. When the MBE system contains a load-lock system in which the mounting block and processed substrate are remotely exchanged with a fresh mounting block and substrate, it is most convenient to have the termocouple come into spring driven contact with the mounting block as it is loaded onto the substrate heater station in the work chamber. Another method of measuring the substrate temperature involves the use of either an optical or infra-red pyrometer (17). The emissivity and window corrections which must be accounted for complicate this method significantly.

3.3 The furnaces, shutters, and baffles

It is the function of the evaporation sources to provide stable, ultrahigh purity molecular beams of the appropriate intensity and uniformity. The intensities of the molecular beams are determined primarily by the vapor pressures of the source materials and secondarily by the evaporation geometry. The expression for the flux angular density distribution (flux/unit solid angle $d\omega$) of material evaporating from a crucible of diameter $2r$ may be written(18,19).

$$\frac{d^2N}{dtd\omega} = 1.11 \times 10^{22} \times \pi r^2 \times \frac{P(T)}{(MT)^{1/2}} \times A(\Theta) \tag{3}$$

where Θ is measured from the crucible axis. The angular term, $A(\Theta)$, for several melt levels, ℓ (measured from the crucible lip), is shown in Fig. 6(19). For a given crucible-substrate geometry the flux density distribution (flux per unit substrate area) can be derived from Eq. (3). For a substrate normal to the crucible axis and a distance L from the crucible lip, the flux density at $\Theta = 0$ may be written(18).

$$\frac{d^2N}{dtds} = 1.11 \times 10^{22} \times \frac{\pi r^2}{L^2} \frac{P(T)}{(MT)^{1/2}}. \tag{4}$$

The vapor pressure vs temperature relationship for the source material can be empirically expressed as

$$\log P(T) \approx \frac{A}{T} + B \log T - C \tag{5}$$

where A, B, and C are constants specific to the element. For Ga, for example,(20)

$$P(T) = 10^{\left[-\frac{11021.9}{T} + 7 \log T - 15.42\right]}. \tag{6}$$

For $L = 12.7$ cm, $r = 0.64$ cm and $T = 1236$ K; $P = 2.03 \times 10^{-3}$ torr and the flux density at the substrate is 6.14 x 10^{14} Ga atoms cm^{-2} s^{-1} resulting in a growth rate of ~ 1 μm h^{-1} of GaAs.

The uniformity of the films grown depends primarily on the uniformities of the molecular beams across the substrate which in turn depend on $d^2N/dtd\omega$ as well as the geometrical relationship between the sources and the substrate. In the absence of wall wetting, Fig. 6 demonstrates that a crucible at uniform temperature acts as a colimator, restricting the flux to progressively more sharply peaked angular distributions as the charge level falls. It is immediately apparent that values based on the cosine distribution associated with true Knudsen cells are far from those obtained with typical crucibles encountered in actual systems which yield progressively greater nonuniformity in the films as the carge is depleted. Knudsen cells can be used where the increase in T needed to compensate the much smaller r value in Eq. (3) is not detrimental. Knudsen cells are also considerably more difficult to outgas, although the introduction of the load-lock has greatly reduced the impact of this problem. It has been shown that properly designed crucible nozzles can achieve collimation of the molecular beams in a predicatable pattern making future improvements in uniformity a matter of crucible engineering(21).

The lifetime of a furnace charge depends upon the required flux density at the substrate, the angular distribution of the flux, and the crucible to substrate distance and orientation. The more collimated the flux angular distribution, the more economical the use of material. However, without specially designed crucible nozzles, economical use of source material and good flux uniformity over the substrate are mutually antagonistic and the emphasis is generally placed on uniformity.

A semiquantitative description of the geometrical effects on film uniformity and charge lifetime can be obtained from an examination of a hypothetical MBE system similar to Fig. 1 consisting of two crucibles, each with a diameter of 1.27 cm and length of 6.4 cm (volume = 8 cm^3) located adjacent to each other. The axes of these two crucibles run through the center of a 2.54 cm diameter substrate wafer located 12.7 cm from the crucible lips. Initially, one crucible is completely filled with Ga and the other with As. For a GaAs film growth rate of 1 μm h^{-1} the initial Ga depletion rate ($\ell/r = 0$) is $\sim 2 \times 10^{-2}$ cm^3 h^{-1}. The accepted As/Ga ratio at the substrate for growth of device quality films is in the range of 5-10. For a ratio of 10, the initial As depletion rate ($\ell/r = 0$) is ~ 0.24 cm^3 h^{-1} and drops to ~ 0.05 cm^3 h^{-1} at exhaustion ($\ell/r = 10$). The average As depletion rate is roughly 0.1 cm^3 h^{-1}, and a full As charge will therefore be exhausted after approximately 80 h or 80 μm of film growth. Just prior to As exhaustion, the Ga charge is approximately 18% depleted ($\ell/r \approx 1.8$), the variation in Ga flux density from the center to the edge of the substrate is $\sim 8\%$, and the As/Ga ratio varies from 10 in the center of the substrate to ~ 8 at the edge of the wafer. Initially, the variation in Ga and As flux density over the substrate is less than 2%. It is also obvious from these calculations that the use of large crucibles for the Group V elements results in long charge lifetimes making the load-lock extremely effective in a properly designed MBE system. The system described in Figs. 2 and 3 accepts eight furnaces simultaneously. Where needed, multiple furnaces can be used for a single element.

With the exception of Si MBE, the resistively heated furnace has served as the "work horse" for MBE. In a typical system each furnace is mounted on its own vacuum flange which contains eletrical heater power feedthroughs and temperature compensated thermocouple feedthroughs. A cut-away sketch of one of the furnaces used in the MBE system of Figs. 2 and 3 is shown in Fig. 7. Where alternating current is used for heating, it is especially important that the furnace heater windings be noninductively wound to prevent distortion of the RED pattern by stray magnetic fields. The furnace thermocouple must physically contact the crucible to provide accurate crucible temperature measurements, particularly at low temperature.

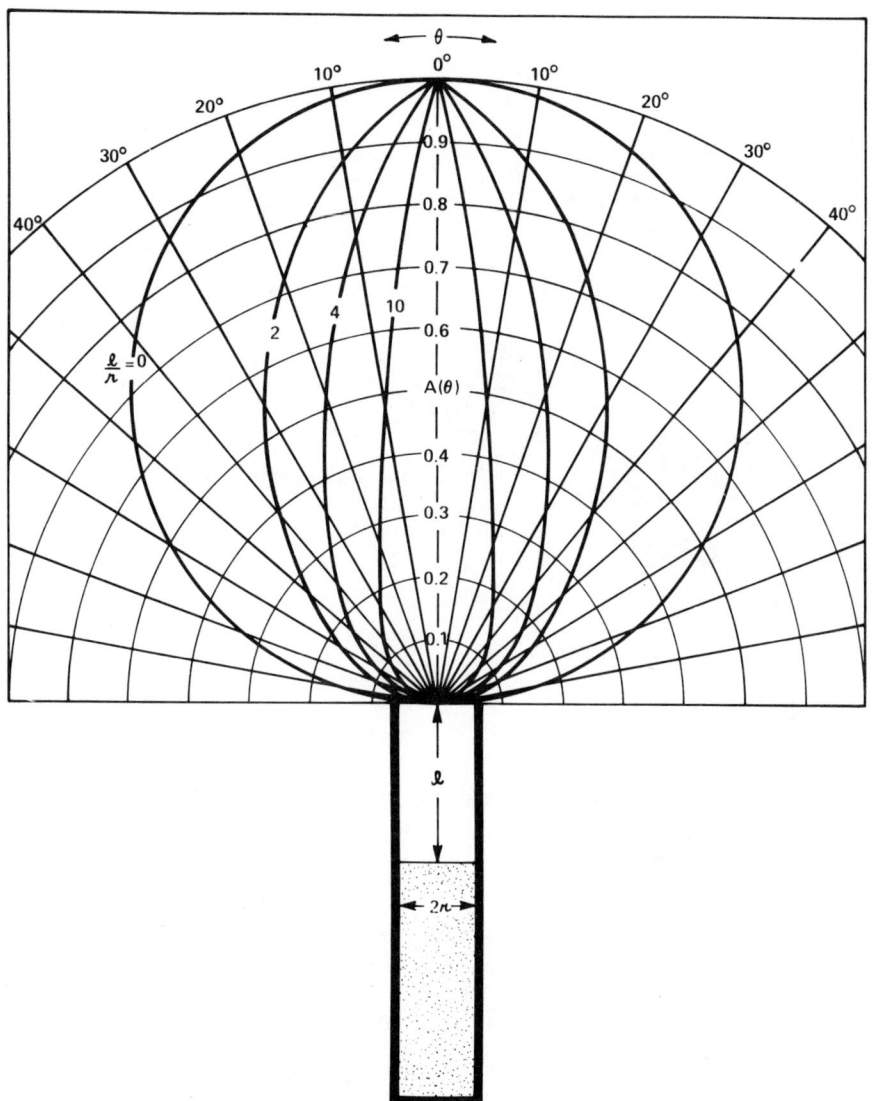

Fig. 6. The angular term, $A(\theta)$, of the flux angular density distribution $d^2N/dtd\omega$ (flux per unit solid angle) is shown for several melt levels l in a crucible of diameter $2r$. (From Ref. 19)

$$\frac{d^2N}{dtd\omega} = 1.11 \times 10^{22} \times \pi/r^2 \times \frac{P(T)}{(MT)^{1/2}} \times A(\theta)$$

Approximate furnace operating temperatures range from a low of $\sim 350°$ C for $P(22)$ to a high of $\sim 1300°$ C for Al(23). The furnaces and crucibles must be fabricated of non-reactive, refractory materials to withstand these high temperatures without contributing themselves to the molecular beams. Tantalum is most commonly used in furnace construction and furnace thermocouples are generally fabricated from tungsten-

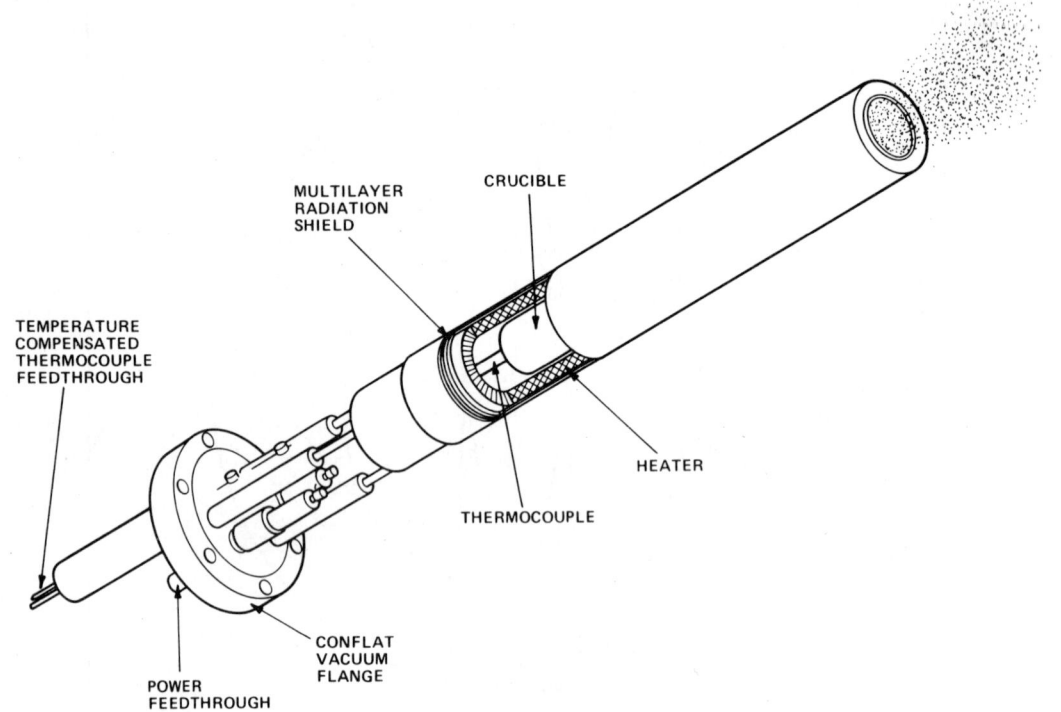

Fig. 7. A cut-away view of the MBE furnace used in the MBE system shown in Fig. 2. (Courtesy of Varian Associates, Inc.)

rhenium alloys. Both graphite and pyrolytic boron nitride (PBN) have been used as crucible materials. Graphite possesses considerable advantage over PBN with respect to cost and machinability, but it is more difficult to outgas and is far more reactive. Graphite plays a significant role in CO generation and will not contain molten Al at high temperature. PBN is therefore becoming the material of choice, particularly at high temperatures with reactive materials. Self-crucible evaporation techniques are required for those source materials which exhibit adequate vapor pressures only at temperatures in excess of the practical limits of available, high purity crucibles or which are too reactive for these crucibles. Electron-beam evaporation of Si is the most prominant example. In carefully designed MBE systems, device quality epitaxial Si films have been grown by electron-beam evaporation(15-17,24).

Multiple radiation shields are integrated into each furnace to improve the temperature uniformity and thermal efficiency. In addition to dramatically reducing demands on the furnace power supply, the shields decrease outgassing due to radiative heating of the system walls. Even with the integral thermal radiation shielding, however, radiation heat transfer is not negligible at high temperatures. Furthermore, chemical cross contamination can occur between adjacent furnaces, even with the shutters closed. Both of these problems are simultaneously minimized with a baffle arrangement which thermally and chemically isolates each furnace and its respective source shutter in its own liquid cooled baffle enclosure. Where LN_2 cooling of these baffles under significant heat load is attempted, precautions must be taken to avoid

N_2 gas pocket formation which reduces localized cooling. Periodic bursts of warm N_2 gas and sudden increases in radiation heat load on LN_2-cooled surfaces can release previously pumped weakly bound species. LN_2 cryopumping in the vicinity of the substrate surface has, however, come to be generally regarded as desirable for reduction of the partial pressures of deleterious gases in the growth region.

The "horizontal" source geometry shown in Fig. 3 is advantageous in minimizing contamination of the furnace charges by flakes falling from deposits which build up on the system walls and shutters. The upper furnaces, whose crucibles face downward, require inverted insert crucibles within the main crucibles to retain the charge. In the case of large furnaces, reasonably sized insert crucibles are possible and the inserts are filled with those materials which are used in small quantities, such as dopants.

3.4 <u>Epitaxy control</u>

To enable reproducible growth, accurate, stable control of the substrate temperature and the relative and absolute molecular beam flux densities at the substrate must be achieved.

In the control of substrate temperature, thermocouple feedback to a rate-proportional controller is straightforward, particularly for substrates mounted by the liquid metal technique. Stability of better than $\pm 1°C$ and accuracy and reproducibility within $\pm 10°C$ are easily achieved and appear quite adequate.

Control of the molecular beam flux is more complex and has been approached both directly and indirectly. The direct control approach involves the use of a beam flux detector (e.g. quartz crystal rate monitor,(13,14,24), ion gauge(24) or quadrupole mass spectrometer (QMS)(25) in real-time flux feedback control of the furnace temperature or power. The most prevalent of these schemes involves the QMS due to its charge/mass discrimination and rapid response. Background nullification has been achieved using chopped beam detection schemes(25). Three major difficulties exist with the direct approach. First, the change in the angular distribution of the flux with variation in charge level described in Fig. 6 results in a significant and continuous variation in the ratio of the flux density at the substrate to that at the detector (e.g. the QMS). Second, the dopant fluxes are so low that detection, let alone control, of the dopant beams is extremely difficult by any of the techniques, especially at low doping levels. Third, even with background nullification, the QMS is subject to significant variation in absolute and relative flux sensitivity due to multiplier gain drift and ionizer space-charge effects (20-50% variation may occur in some cases).

In the indirect approach, the crucible temperature is controlled by thermocouple feedback to a rate-proportional controller. The temperature related stability of the Ga molecular beam flux can be closely approximated from Eq. (6) as

$$\frac{dP(T)/dT}{P(T)} = 2.3 \left[\frac{11021.9}{T^2} + \frac{3.04}{T} \right] . \qquad (7)$$

Thus for a Ga furnace operating at 1236 K (growth rate of $\sim 1~\mu m~h^{-1}$)

$$\frac{dP}{P} = 0.022~dT . \qquad (8)$$

A 1 K variation in temperature therefore results in a 2.2% variation in molecular beam flux. Since a stability of better than $\pm 1/2$ K is easily achieved, flux stability of better than $\pm 1.1\%$ is easily achieved. Furthermore, Fig. 6 illustrates that if the charge temperature remains constant, the central flux density ($\theta = 0°$) remains unchanged even though the peripheral flux density ($|\theta| > 0°$) decreases (along with film uniformity). However, in full crucibles with large orifices, which are generally encountered in MBE systems, the charge temperature does in fact change as the level falls, and a periodic re-normalization scheme is needed if the level is allowed to change significantly. Inasmuch as Ga usage is small, re-normalization of the Ga molecular beam is needed infrequently. One of the simplest and most successful re-normalization techniques involves achieving a specified pressure reading for each molecular beam at the precise substrate growth position. This can be achieved by mounting an ion gauge on the substrate heater carousel which may be rotated directly into the growth position.

In some material systems it is desirable to reduce relative molecular beam fluctuations by as much as an order of magnitude over that which can normally be achieved with separate source furnaces. For a given value of absolute crucible temperature control, fluctuations in the layer composition can be considerably reduced by integrating two separate effusion cells into a single crucible body(26). By making the crucible of materials with significant cross section and thermal conductivity, the crucible is essentially isothermal and the beam fluctuations from the separate cells are in phase. Relative flux densities are then determined by the radio of individual cell orifice diameters (26).

3.5 In situ substrate preparation

In the MBE system described in Figs. 2 and 3 the substrate can be rotated from the growth position to any of a number of cleaning or analysis positions without change in temperature. Three *in situ* cleaning methods have been developed: heating in UHV of properly prepared substrates(1,27) heating in 1 x 10^{-6} torr of water vapor or oxygen, (3,28) and sputtering with low energy (< 500 eV) inert gas ions followed by annealing (1,3). The substrate cleaning ion gun shown in Fig. 3 is orientated normal to the substrate and the ion beam can be rastered to achieve uniform etching over large substrates. The method actually chosen depends upon the materials and contaminants involved as well as the device structure being grown.

If the substrate requires more than a very brief sputter or heat cleaning, the throughput of the growth system can be increased by the inclusion of a pre-preparation chamber between the loading chamber and the work chamber. Lengthy procedures, such as extended sputtering or mask evaporation, or processing potentially deleterious to the work chamber environment, such as heating in oxygen or water vapor to remove carbon, can be performed in parallel with growth operations. UHV techniques should prevail for the preparation and transfer system since the surface, once cleaned, should not be significantly disturbed prior to initiation of growth.

3.6. Surface analytical instrumentation

In the growth of device quality epitaxial semiconductor films, by far the most sensitive, and, in fact, relevant, indications of correct growth parameters are the electro-optical and electronic transport properties of the films. These measurements (e.g. photoluminescence and Hall mobility, respectively) are performed external to the MBE system and are sensitive to contaminant and defect concentrations well below the detection capabilities of surface analysis instrumentation such as Auger electron spectroscopy (AES) and secondary ion mass spectroscopy (SIMS). The final word, of course, is the actual performance of the material in the desired device. The dramatic increase in throughput made possible by the use of load-locks has greatly expedited variation of the growth parameters to optimize the basic semi-conductor material parameters.

In the development of growth processes, however, surface-sensitive analytical equipment has played an extremely valuable role. One example is the identification, with AES, of surface contamination responsible for growth defect nucleation(1).

Table 1 lists those surface analysis techniques which are most applicable to MBE. Since growth by MBE is a kinetic process which occurs far from thermodynamic equilibrium, the ability to examine the surface *in situ* during growth is very valuable. RED provides this capability since the electron gun and the phosphor screen which displays the diffraction pattern are well removed from the growth region and hence do not interfere with the molecular beams. The diffraction pattern directly provides information regarding the surface crystal structure and morphology, and, in conjunction with Auger electron spectroscopy studies, can be correlated with surface chemistry. It is important to note, however, that growth of films for device applications should be carried out without the RED gun operating since electrons impinging on the surface during growth can greatly enhance the sticking coefficients of residual gases and hence increase the impurity levels in the films.

Table 1

TECHNIQUE	PRIMARY FUNCTION
Mass Spectroscopy	Molecular Beam Flux Analysis
	Residual Gas Analysis
RED (HEED)	Surface Crystal Structure (Unit Cell Symmetry and Dimensions)
	Surface Roughness
Auger Spectroscopy	Surface Analysis (Spatial resolution $< 5\ \mu$m)

	— Stoichiometry
	    — Chemical States
	    — Contamination

	Depth Profiles (Depth resolution <50 Å)
	— Matrix Composition Profiles — Dopant Profiles (in heavily doped materials only)
SIMS	Surface Analysis (Spatial resolution $\gtrsim 100\ \mu$m)
	— Contamination (Extremely sensitive for certain elements) — Chemical States
	Depth Profiles (Depth resolution <50 Å)
	— Matrix Composition Profiles — Dopant Profiles (Extremely sensitive for certain elements)
Optical Reflection	Surface Roughness

The MBE system described in these pages is a sophisticated and complex instrument. It should be noted, however, that much of the analytical equipment is not essential once a specific growth process is well developed. As a result, the complexity, and hence the cost, of a production MBE system may be reduced significantly from that of a research and/or developmentsystem such as that described here.

REFERENCES

1. A. Y. Cho and J. R. Arthur, *Prog. Solid State Chem.* 10, 157 (1975).
2. L. L. Chang and R. Ludeke, In *Epitaxial Growth, Part A*, ed. J. W. Matthews, Academic Press, New York, 1975, p. 37.
3. B. A. Joyce and C. J. Foxen, Inst. Phys. Conf. Ser. No. 32 (1977) p. 17.
4. P. E. Luscher, *Solid State Technol.* 20, 43 (1977).
5. L. L. Chang, L. Esaki, W. E. Howard, R. Ludeke, and G. Schul, *J. Vac. Sci. Technol.* 10, 655 (1973).
6. A. C. Gossard, P. M. Petroff, W. Weigmann, R. Dingle, and A. Savage, *Appl. Phys. Lett.* 29, 323 (1976).
7. P. M. Petroff, *J. Vac. Sci. Technol.* 14, 973 (1977).
8. D. Smith, this issue.
9. J. R. Arthur, *J. Appl. Phys.* 39, 4032 (1968).
10. S. Dushman, *Scientific Foundations of Vacuum Technique*, Wiley, New York, 1949.
11. G. E. Becker, *J. Vac. Sci. Technol.* 14, 640 (1977).
12. D. W. Covington and E. L. Meeks, Reported at MBE '78, Paris, France (1978).
13. D. L. Smith and V. Y. Pickhardt, *J. appl. Phys.* 46, 2366 (1975).
14. E. Kasper, H. J. Herzog, and H. Kibble, *Appl. Phys.* 8, 199 (1975).
15. U. Konig and H. Kibble, Reported at MBE '78, Paris, France (1978).
16. G. E. Becker and J. C. Bean, *J. appl. Phys.* 48, 3395 (1977).
17. Y. Ota, *J. Electrochem. Soc., Solid-State Science Technol.* 124, 1795 (1977).
18. M. Knudsen, *Ann. Physik*, 48, 1113 (1915).
19. B. B. Dayton, *1956 Vacuum Symposium Transactions*, Committee on Vacuum Techniques, Boston pp. 5-11.
20. Taken from R. E. Honig and D. A. Kramer, *RCA Review*, 30, 285 (1969).
21. L. Y. L. Shen, *J. Vac. Sci. Technol.* 15, 10 (1978).
22. J. H. McFee, B. I. Miller, and K. J. Bachmann, *J. Electrochem. Soc.* 124, 259 (1977).
23. A. Y. Cho, R. W. Dixon, H. C. Casey Jr., and R. L. Hartmann, *Appl. Phys. Lett.* 28, 501 (1976).
24. Y. Matsushima, Y. Hirofuji, S. Gonda, S. Mukai, and M. Kimata, *Jap J. appl. Phys.* 15, 2321 (1976).
25. L. L. Chang, L. Esaki, W. E. Howard, and R. Ludeke, *J. Vac. Sci. Technol.* 10, 11 (1973).
26. H. Holloway, D. K. Hohneke, R. L. Crawley, and E. Wilkes, *J. Vac. Sci. Technol.* 7, 586 (1970).
27. A. Y. Cho and F. K. Reinhart, *J. appl. Phys.* 45, 1812 (1974).
28. C. E. C. Wood, *Appl. Phys. Lett.* 29, 746 (1976).

THE AUTHORS

Paul E. Luscher

Douglas M. Collins

Paul E. Luscher is manager of the MBE Advanced Development Department of Varian Associate's Surface Technology Operation, Palo Alto, California. He joined Varian's Surface Analytical Instrument's Advanced Development Department in 1975, where he was primarily concerned with the application of sputter Auger depth-profiling to semiconductor device interfaces. He is presently acting manager of this department.

Prior to joining Varian Associates, his principal research activities were in the study of adsorption and reaction on solid surfaces. His major contributions in this area were in the elucidation of mechanisms responsible for structure in the inelastic electron scattering spectrum from tungsten at low energy, and the application of inelastic electron scattering to the study of the substrate-adsorbate electronic structure. His work in instrument design includes UHV-compatible dispersive and non-dispersive electron and ion optics for surface research applications.

He received his B.S. in Science Engineering in 1966, and his Ph.D. in Materials Science in 1975, from Northwestern University, Evanston, Illinois. From 1970 to 1975, he held a research assistantship at the Coordinated Science Laboratory, University of Illinois at Urbana-Champaign.

Douglas M. Collins received his B.S. degree in Electrical Engineering from Montana State University in 1971; he receive his M.S. degree (1972) and Ph.D. (1977) in Electrical Engineering from Stanford University. His thesis research involved ultraviolet photoelectron spectroscopy, thermal desorption spectroscopy, and Auger electron spectroscopy studies of gas adsorption on Pt. Since June 1977 he has been involved with the III-V compound molecular beam epitaxy program in the Corporate Research Laboratories of Varian Associates.

Dr. Collins is a member of Tau Beta Pi, Sigma Xi, the American Physical Society, the American Vacuum Society, and the Institute of Electrical and Electronics Engineers.

NONSTOICHIOMETRY AND CARRIER CONCENTRATION CONTROL IN MBE OF COMPOUND SEMICONDUCTORS.

Donald L. Smith

The Perkin-Elmer Corporation
Norwalk, CT 06856, U.S.A.

(Submitted May 1978)

ABSTRACT

All compound semiconductors remain single-phase over some small range of deviation from stoichiometric proportions. Outside this range precipitates occur. Within it, native point defects including vacancies, interstitials and antisites are generated; these act as shallow or deep acceptors and donors and thus profoundly affect the electronic properties of the crystal. The equilibrium partial pressure of N_2 over the compound MN typically increases orders of magnitude as stoichiometry deviation varies from the M-rich to the N-rich phase boundary. Although MBE is not an equilibrium process, film stoichiometry deviation is widely adjustable with the N_2/M impingement rate ratio during growth. Closer control of this ratio is possible for the II-VI and IV-VI compounds by using the binary compounds as evaporation sources rather than the constituent elements, since the II-VI compounds evaporate dissociatively and congruently (the elements being more volatile than the compounds) and the IV-VI compounds evaporate mostly as MN. The III-V compounds evaporate dissociatively and noncongruently, since the III elements are less volatile. At growth temperatures just below the re-evaporation temperatures of the compounds, where highest crystal quality is generally obtained, the composition of II-VI compound films is self-regulating within the single-phase field because of the high volatility of the elements. A small fraction of excess impinging Te_2 is incorporated into ZnTe, however, probably in the form of Zn vacancies, since these acceptor defects are thought to be responsible for the high p-conductivity of pure ZnTe. GaAs grown with high As_2/Ga impingement flux ratio contains a photoluminescent deep defect level which is absent with lower As_2/Ga. In addition, the amphoteric impurity dopant Ge may be made to move from Ga (donor) sites to As (acceptor) sites by decreasing As_2/Ga, but usually with the appearance of a Ga precipitate. In the IV-VI compounds, stoichiometry deviation produces only shallow vacancy and interstitial defect levels, which act as dopants. Defect doping is typically adjustable from 10^{19} p cm^{-3} using excess VI flux to 10^{17} n cm^{-3} using excess IV flux, but only 0.1-1% of the excess elements are incorporated into the growing film. Most excess Se and Te re-evaporate and most excess Pb precipitates at the growth surface. The M-substitutional impurity dopants Tl (p-type) and Bi (n-type) are more suitable for IV-VI MBE, since nearly all impinging flux is actively incorporated and since higher doping levels (10^{19}-10^{20} cm^{-3}) are achievable. At the higher doping levels, compensation occurs in some cases as dopant begins to be incorporated on VI sites also, but this effect can be reduced by the use of excess impinging VI. Alternatively, compound dopants such as Bi_2Te_3, which evaporates molecularly, can be employed.

1. THERMODYNAMIC FOUNDATIONS

This review deals with the control of stoichiometry deviation in MBE growth of compound semiconductors, and the effects of such deviation on the electronic properties of the grown films. It will deal specifically with binary compounds of the form $M_{1-x}N_x$, where x ≈ 0.5, formed from one metal (M) and one non-metal (N) atom. Examples will be given from those of the group IIB-VIA (such as ZnTe), IIIA-VA (such as GaAs) and IVA-VIA (such as PbTe) compounds with which we have had experience in our laboratory. Like all compounds, these crystals can accommodate a certain degree of deviation, Δx, from the stoichiometric proportion, x = 0.5, before a second phase appears in the form of a precipitate. It is the object of any crystal growth technique to avoid such precipitates and also to manipulate the stoichiometry deviation within the single-phase field to optimize the electronic properties of the crystal. The discussion below will develop some guidelines for achieving these objectives in MBE and will illustrate the control possible for various compounds.

Stoichiometry deviation is accommodated by a crystal lattice by the generation of native point defects. The three basic types, those involving only one lattice site each, are listed in Table 1 along with their expected electrical behavior (1). There are also various more complex and less common point defects involving more than one lattice site which will not be discussed here. An acceptor defect can be electrically neutral or can accept an electron from the lattice and become negatively charged, depending on the position of Fermi level relative to the energy level of the defect in the energy band diagram. Conversely, a donor defect can be neutral or positively charged. The effect of charged defects on the electronic properties of the crystal depends on the defect energy level relative to the band edges. A donor whose energy level is close to the conduction band edge (a shallow donor) easily loses an electron to the conduction band and thus acts as an n-dopant. Similarly, a shallow acceptor point defect acts as a p-dopant. Defects with energy levels deeper within the energy gap act instead as majority carrier traps or as recombination centers. Both shallow and deep defects are commonly observed in the compounds MN. It should also be noted that N-rich material in which either vacancies, V_M, or interstitials, N_i, are predominant, exhibits acceptor behavior; whereas if antisites, N_M, are predominant, donor behavior is observed.

The nonstoichiometry behavior of the compound MN may be described thermodynamically in terms of the T-x projection of the equilibrium phase diagram, as shown in generalized form in Fig. 1a. Here, the nonmetal fraction x is plotted vs equilibration temperature T. Below the melting point of MN, there is some Δx over which the single solid phase $MN(s)$ is stable. Beyond the edge of this region (the solidus boundary) a second phase appears in the form of an M or N precipitate, so that the solidus boundary represents the solubility limit of point defects. The width, Δx, of the single-phase existence region is very small, typically 10^{-3} - 10^{-6}. Since there is a certain activation energy E_a associated with the creation of such defects in a perfect lattice, the solubility limit generally increases exponentially with T as in the cases of other activated processes, following an equation of the form (1):

$$\log (V_M) \sim - E_a/KT(K), \qquad (1)$$

Table 1. Basic native defects in compound semiconductors MN

		Electric behavior
Vacancies	V_M	acceptor
	V_N	donor
Interstitials	M_i	donor
	N_i	acceptor
Antisites	M_N*	acceptor
	N_M	donor

* Metal atom on nonmetal site.

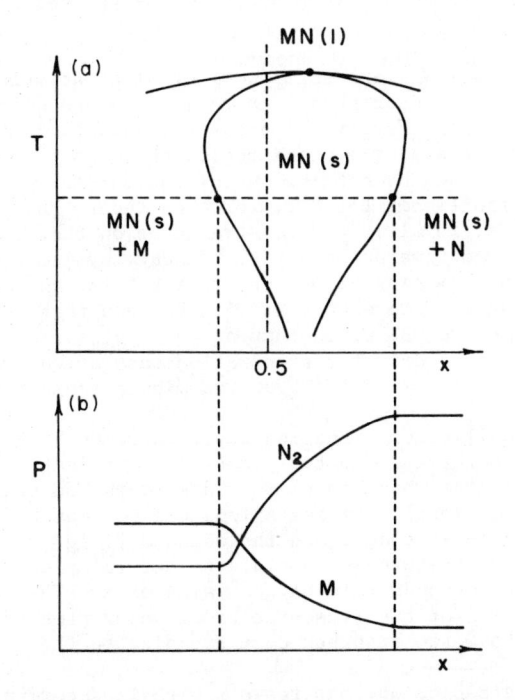

Fig. 1. General form of phase equilibria for nonstoichiometric binary semiconductors $M_{1-x}N_x$; (a) = T-x projection; (b) = P-x projection.

at least in the region well away from the melting point. This so-called "retrograde"' solubility is often a source of difficulty in crystal growth; for if upon cooling a nonstoichiometric crystal, the solidus is crossed, microprecipitates may be formed throughout the bulk. Note that the solidus field does not necessarily even include the stoichiometric composition.

There is also a vapor phase in equilibrium with the condensed matter at any point on the T-x diagram, and the equilibrium partial pressures of the various vapor species will be functions of T and x. As it turns out, the major vapor species over the compound semiconductors MN are in general MN, M, and N_2. Since nonstoichiometric MN is, in effect, a dilute solid solution of excess M or N in $M_{0.5}N_{0.5}$ the equilibrium partial pressure of N_2 would be expected to increase with concentration of dissolved N as the N-rich solidus boundary is approached, as it would be expected to do for any dilute solution. In fact, the increase is typically many orders of magnitude from boundary to boundary of the existence region, with a corresponding decrease in P_M such that the equilibrium constant

$$K = P_M^2 \, P_{N_2} \qquad (2)$$

is held fixed. This situation is illustrated in generalized form in Fig. 1b for a given T shown by the dashed line in Fig. 1a. When the solidus boundaries are reached, the second condensed phase begins to form, and the partial pressures are then fixed. This is in accordance with Gibb's phase rule:

$$\text{(degrees of freedom)} = \text{(components)} + 2 - \text{(phases)}; \qquad (3)$$

for with two components (MN + excess M or N; or alternatively $M + N$), and with two condensed phases plus the vapor phase, there is only one degree of freedom: once T is specified, so are all the partial pressures as well as x for each phase. On the other hand, within the single-phase region there is a second degree of freedom, so that x may be determined by setting one of the partial pressures. This last effect is the most significant for MBE, for even though MBE cannot be considered an equilibrium process, it will be demonstrated below that the composition of the grown film may be driven toward the N-rich boundary by increasing the ratio N_2/M of molecular beam fluxes impinging on the growing film, and towards the M-rich boundary by decreasing it. Moreover, it will be seen that the equilibrium phase diagram applies further to MBE to the extent that the solidus boundaries are indeed encountered in the form of defect concentration saturation or of precipitates on the growth surface if N_2/M is varied over too wide a range.

In visualizing the equilibrium situation represented by Figs. 1a and 1b, it is helpful to remember that these are two-dimensional projections of the three-dimensional P-T-x phase diagram which completely describes the nonstoichiometric compound MN. In the three-dimensional representation, the solidus boundaries are three-dimensional curves in P-T-x space, and the solidus field is a sheet connecting the curves. Fixing the temperature and one of the partial pressures determines a point on the sheet which corresponds to a unique value of x. The third projection, P vs T, is the familiar form for the presentation of vapor pressure data, and for the compound MN it will have two branches corresponding to the two solidus boundaries.

In MBE growth, the variables one has to work with in determining the composition x of the grown film are the substrate temperature and the impingement fluxes of the evaporants on the substrate. In practice, the substrate temperature is the less significant of the two for purposes of stoichiometry control, for it is usually constrained within a narrow range of 100°C or so, bounded by re-evaporation of the film on the upper end and by loss of crystal quality on the lower end. Therefore, given a predetermined substrate temperature, the objective is to adjust N_2/M in the impinging beam to first avoid precipitates and to further regulate x within the single-phase region to optimize the electronic properties of the film. Impingement flux ratio control requires knowledge of the evaporation mechanisms of the compound semiconductors if these are to be used as source materials rather than the elements. Finally, the relationship between the impingement flux ratio and the composition of the film itself is determined by the relative condensation and incorporation rates of the various impinging species on the growing film.

The principal evaporation mechanisms for the three classes of compounds of concern here are, by way of specific example:

$$\text{II-VI's:} \quad ZnTe(s) \rightarrow Zn(g) + 1/2\ Te_2(g) \qquad (4)$$

$$\text{III-V's:} \quad GaAs(s) \rightarrow Ga(l) + 1/2\ As_2(g) \qquad (5)$$

$$\text{IV-VI's:} \quad PbTe(s) \rightarrow PbTe(g). \qquad (6)$$

In the case of the II-VI compounds, the vapor pressures of the constituent elements over themselves are much higher than those over the compounds, so no condensed phase is formed upon dissociation of the compounds. As evaporation proceeds, the solid composition x adjusts itself to the point within the solidus field at which the atomic evaporation rate of N_2 equals that of M (near where the curves cross on Fig. 1b), and evaporation then proceeds congruently. The evaporation of a II-VI compound thus results in a stoichiometric N_2/M impingement rate ratio on the growing film. Of course, this situation only results in a stoichiometric film if the incorporation rates of M and N_2 are exactly equal. In general, a supplementary evaporation source of one or the other element would be required to adjust film stoichiometry. Alternatively, the pure elements alone could be used as sources, but the exponential dependence of P on T makes N_2/M ratio control much more difficult in this instance. Vapor pressures of II-VI compounds have been extensively investigated (2). III-V compound evaporation also proceeds dissociatively (eq. 5), with a small amount of As_4 also being formed. However, Ga is less volatile than either As_2 or P_2 even at the Ga-rich solidus (that is, the curves on Fig. 1b do not cross), at least at temperatures high enough for useful evaporation rates, so that evaporation is always noncongruent and N_2-rich. Most impinging Ga must therefore be supplied from a separate source of pure Ga, making N_2/M ratio control much more difficult than for the II-VI compounds. Vapor pressures over both GaAs (3) and GaP (4) have been reported. The IV-VI compounds evaporate mostly as MN molecules (eq. 6), but about 14% dissociates into ($\frac{1}{2}M + N_2$) in the cases of PbTe (5) and SnTe (6). While the dissociated fraction of PbTe evaporates congruently, that of SnTe does not because of the low volatility of Sn. Additional discussion of the evaporation of these three groups of compounds has appeared in a recent general review of MBE (7).

Given that the impingement rate ratio N_2/M can be adjusted at will by using a binary compound evaporation source supplemented by an elemental source or by using two elemental sources, it remains to determine the relationship between this ratio and the composition of the grown film. Incorporation into the film involves, first, condensation of the element on the surface, and then reaction with the surface lattice. A first-order picture of the relative condensation probabilities of the various species involved may be obtained by comparing the temperatures at which their vapor pressures reach 10^{-6} torr. This is a useful pressure at which to make such comparisons, because it corresponds to a surface evaporation or condensation rate of about 1 monolayer sec^{-1} or about 1μm hr^{-1}, which is a typical MBE growth rate. Species whose vapor pressures pass 10^{-6} torr at temperatures well below the desired growth temperature would be expected to have very low condensation probabilities on the growing film. Those which reach 10^{-6} torr well above the growth temperature will condense on the growth surface, after which they might either become incorporated into the growing film by reaction with the surface lattice or might remain on the surface as a precipitate. Both effects are observed in MBE. Figure 2 shows the 10^{-6} torr temperatures arranged by period for the elements from groups II through VI (from Ref. 8) and for the three groups of binary compounds under discussion. Compound evaporation rates were taken at the congruently evaporating composition, or, if none, at the metal-rich solidus where P_N is a minimum. It should be borne in mind in examination of Fig. 2 that the optimum MBE growth temperature is generally just below the temperature at which the compound evaporation rate begins to approach the growth rate, or 50-100°C below the temperatures given in Fig. 2, for a 1 μm hr^{-1} growth rate. The three groups of compounds will be discussed separately below, with reference to Fig. 2 and with the introduction of additional data regarding the dependence of stoichiometry deviation on the N_2/M flux ratio.

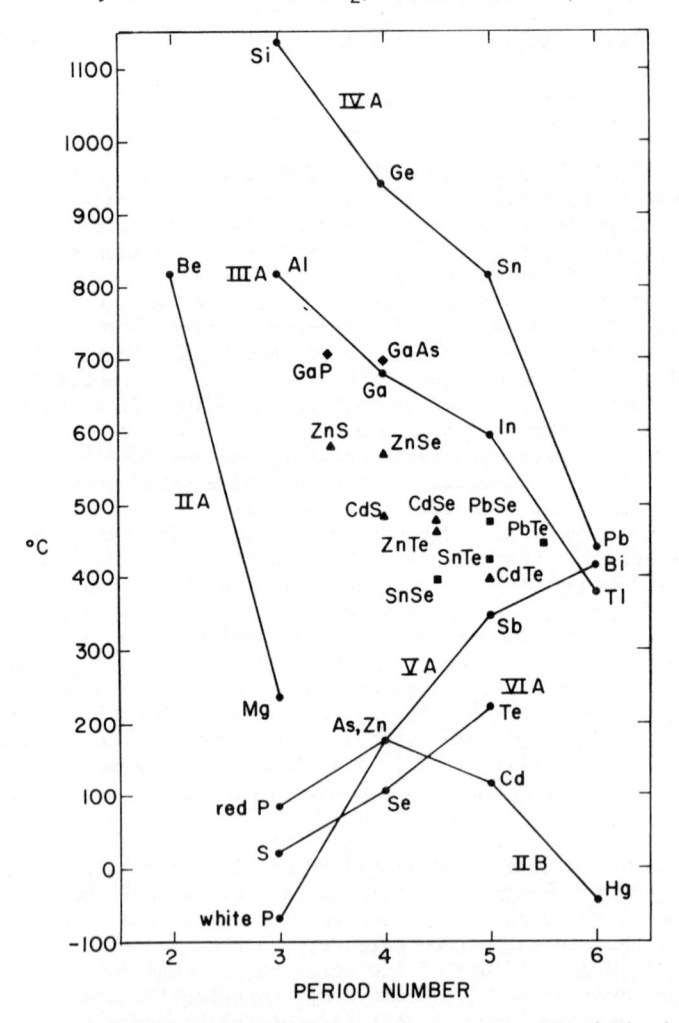

Fig. 2. Temperatures at which evaporation rates of the binary semiconductors and their constituent elements reach 10^{-6} torr (approximately 1 monolayer sec^{-1} or 1 μm hr^{-1} evaporation rate).

2. II-VI COMPOUNDS

Inspection of Fig. 2 reveals that the group IIB and VIA elements are all much more volatile than their compounds, so that one would expect no precipitates of either M or N on the growth surface; that is, film stoichiometry is self-regulating at least to within the solidus boundaries. This self-regulation was demonstrated in our laboratory by growth from elemental evaporation sources of a number of II-VI compounds onto a resonant quartz crystal mass deposition monitor which was held at the growth temperature (9). With M flux held constant, deposition rate G was monitored while N_2 was increased. G increased in proportion to N_2, indicating that all N_2 was condensing, until approximately the stoichiometric flux ratio, after which G ceased to increase with additional N_2 flux, indicating that excess N_2 was re-evaporating. The same result was obtained when M was increased with N_2 held constant. Although the deposit was, of course, polycrystalline, the general features would be expected to be the same for epitaxy: near-unity condensation probability of one element in the presence of excess flux of the other element, and near-zero in deficient flux of the other element.

In the case of ZnTe, a slight slope in G in the excess Te region suggested about 4% incorporation of excess Te (9). This probably corresponds to the generation of V_{Zn}, since this acceptor defect has been suggested as the source of the high p-type conductivity commonly observed in unintentionally-doped bulk ZnTe (10). More recent MBE work has shown that MBE ZnTe is also p-type (11,12). There does not appear to be any MBE work in which carrier concentration was systematically studied as a function of N_2/M flux ratio. But ZnTe annealed in increasing over-pressures of Zn vapor shows a decreasing p-type conductivity (13), so it is quite likely that the carrier concentration of MBE ZnTe could be adjusted with N_2/M flux ratio. Except for CdTe, the extent of the solidus fields are not well known for the II-VI compounds, although the available data has been well reviewed in Ref. 10. The defect structure of CdTe has been studied more recently in Ref. 30.

3. III-V COMPOUNDS

Figure 2 shows that while As and P are much more volatile than GaAs and GaP, P_{Ga} does not reach 10^{-6} torr until above the 600°C growth temperature. Thus, excess N_2 re-evaporates, but excess M remains as a surface precipitate. The kinetics of these processes were studied in detail for GaAs in a classic series of experiments early in the development of MBE (14), and MBE growth of GaAs devices has also been thoroughly reviewed (15). Figure 2 also demonstrates why many of the common GaAs dopants are unusable in MBE: the substitutional dopants from groups IIA and B (p-type) and from group VIA (n-type) all reach 10^{-6} torr at very low temperatures, with the exception of Be. Ionization of Zn, however, has been shown to increase its incorporation rate to a usable level (16).

It is of interest to compare what is known about the extent of the solidus field with the experimental effects of N_2/M ratio adjustment on the electronic properties of the crystal. GaP is one of the few compounds for which direct chemical measurements of stoichiometry deviation have been possible. The complete solidus field has been calculated from these measurements, which were made at the higher temperature (wider) end, and from other available thermodynamic data, with the assumption that only the vacancy defects, V_{Ga} and V_P, were important (17). Figure 3 shows a further extrapolation of these calculated boundaries down to the MBE growth temperature range, made using the log (P) vs $1/T$ relationship (eq. 1). Although it has more recently been suggested (18) that antisite defects might also be important, Fig. 3 nevertheless represents an order-of-magnitude estimate of the solidus boundaries. Figure 3 also presents the Ga-rich solidus boundary of GaAs similarly extrapolated from calculations which assumed both vacancies and antisites (19). Interstitials were omitted due to very high predicted activation energies. Although the As-rich boundary was not reported in a form which could be so extrapolated, the high temperature solidus field was skewed towards the Ga-rich side as the case of GaP. Some V_{As} was also present at the Ga-rich solidus, but Ga_{As} was the dominant defect.

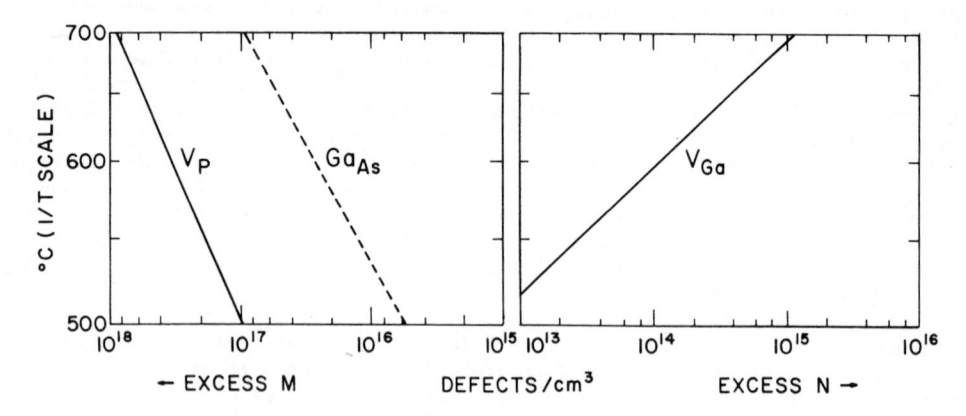

Fig. 3. Calculated solidus boundaries and dominant point defects for GaP(17) and GaAs(19) in the MBE growth temperature range.

Although the degree of stoichiometry deviation does not have a significant effect on the carrier concentration of GaAs or GaP, there is considerable experimental evidence connecting As-rich GaAs with the occurrence of deep levels not present in Ga-rich material, and it has often been assumed that these are due to V_{Ga} defects. Such levels were found in the photoluminescence spectrum of bulk GaAs which was annealed in As-rich vapor (20) as well as in MBE GaAs grown under high As_2/Ga flux (21). Deep-level transient capacitance spectroscopy measurements on CVD GaAs showed a deep level whose concentration increased from 2 to 9×10^{13} cm^{-3} when the AsH_3/GaCl reactant ratio was changed from 1/3 to 3 (32). MBE GaAs is therefore best grown in the narrow range of As_2/Ga flux ratio above which the deep defect level begins to appear and below which Ga precipitates begin to form on the growth surface. There remains the question of why the relatively large concentration of Ga_{As} defects in Ga-rich material remain electrically inactive, if indeed they are present in the quantity which has been calculated.

The N_2/M flux ratio also in general is expected to affect the incorporation of impurity dopants in the growth of any of the compound semiconductors, because it affects the relative surface population of M or N lattice sites into which they can fall. The group IVA dopants have near-unity condensation probabilities on MBE GaAs, as expected from Fig. 2, and both Si and Ge are almost entirely incorporated as active dopants, only Sn being subject to surface segregation. The group IVA dopants are amphoteric, being n-type when substituted on Ga sites and p-type on As sites. In practice, Si and Sn are always predominantly n-type, while Ge is n-type under high As_2/Ga growth conditions and p-type under low As_2/Ga, as it becomes forced into V_{Ga} sites or V_{As} sites, respectively. This effect is shown in Fig. 4, which plots Hall mobility vs As_2/Ga flux ratio for a series of films all grown at 560°C with an impurity concentration of 2×10^{18} Ge cm^{-3}. As_2 and Ga fluxes were measured directly using a room temperature quartz crystal monitor, which we found to have a unity condensation probability for As_2 provided that sufficient Ga flux is simultaneously impinging on the crystal. For As_2/Ga <0.7, Ge is p-type with a mobility equal to good LPE material, indicating that all Ge is entering As sites. The transition to n-Ge is very abrupt at $As_2/Ga = 0.7$, although mobility only reaches 1/2 the value of LPE material, indicating compensation by p-Ge. n-mobility would probably increase with still higher As_2/Ga ratio. The incorporation of Ge in p or n sites has also been correlated with the presence during growth of a "Ga-stabilized" or an "As-stabilized" surface structure as observed by grazing-incidence electron diffraction (22). In practice, it is difficult to control As_2/Ga to generate p-Ge without encountering surface droplets of Ga: on this basis Be or Zn^+ would be a better choice as a p-dopant. It is nevertheless clear that one can control the relative availability of Ga or As sites by adjusting the As_2/Ga flux ratio.

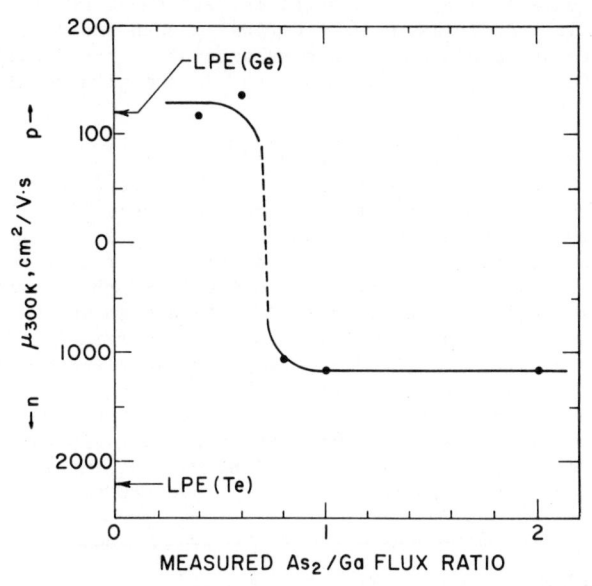

Fig. 4. Doping behavior of Ge in GaAs as a function of As_2/Ga impingement flux ratio during growth: 560°C growth on GaAs(100), 2×10^{18} Ge cm^{-3}.

4. IV-VI COMPOUNDS

The condensation probabilities of M and N_2 for the IV-VI compounds are similar to those for the III-V's (see Fig. 2): most excess VI re-evaporates, while excess IV remains as a precipitate. The M-substitutional impurity dopants Bi (n-type) and Tl (p-type) would also be expected from Fig. 2 to have high condensation probabilities, as in fact they do. Regarding impurity dopants for these compounds, annealing studies of impurity-doped PbTe in Pb-rich or Te-rich atmospheres has shown that, from among the group IIIA, VA, and other appropriate dopant elements, Bi and Tl are the most well-behaved in that their expected behavior as M-substitutional dopants is the least affected by changes in nonstoichiometry caused by the annealing (23). The incorporation as well as the diffusion behavior of these dopants has been extensively studied in our laboratory (24). Elemental evaporation sources have been used; and Bi_2Te_3 has also been used as a Bi source, since it oxidizes less easily upon exposure to air, and since its evaporation as an undissociated molecule presents a basically different chemical situation at the growth surface than the use of elemental Bi.

The solidus field and point defect composition of the IV-VI compounds are much more clearly defined than are those of the II-VI or the III-V compounds. Annealing studies of IV-VI compound wafers in M-rich or N-rich atmosphere has shown that carrier concentration saturates at an n or p level, respectively, which is a function of annealing temperature (25-27). The stoichiometry deviation of the wafers was being shifted in these experiments from the M-rich to the N-rich solidus boundary, and there was an accompanying shift in the dominant point defects from donor-type to acceptor-type. This means that antisite defects cannot be important, since they would behave the opposite way with stoichiometry deviation (see Table 1). Moreover, carrier concentration is no lower at cryogenic temperatures than at room temperature: the carriers so not "freeze out". Thus, the defect structure of the IV-VI compounds consists only of vacancies and interstitials, and these create only shallow donor and acceptor levels. Analysis of self-diffusion data indicates that the relative importance of interstitaials and vacancies depends on the position within the solidus field (28). There is no evidence for deep native point defect levels in the IV-VI compounds.

The solidus boundary data found from annealing studies (26) are plotted in Fig. 5, in terms of carrier concentration rather than absolute defect concentration. While it is generally assumed that each defect produces one mobile charge, theoretical calculations suggest that each may produce two here (29). Reported data at temperatures above the top of Fig. 5 (>600°C) continue to follow a log (n) vs $1/T$(K) relationship, so the short extrapolation to the 400°C MBE growth temperature range is considered very reliable. Solidus field data are shown in Fig. 4 for PbTe and for two ternary alloys of the form $Pb_{1-y}Sn_yN$. The particular ternary compositions shown have energy bandgaps in the 10-15 μm photon wavelength range at cryogenic temperatures, with device applications to photovoltaic infrared detectors and diode lasers. Note that the solidus fields are all much wider than those of the III-V compounds, those for the ternary compounds being especially wide. All are strongly skewed to the p-type, or N-rich side, so that in case of $Pb_{0.8}Sn_{0.2}Te$, even the M-rich boundary is p-type, at least above 540°C.

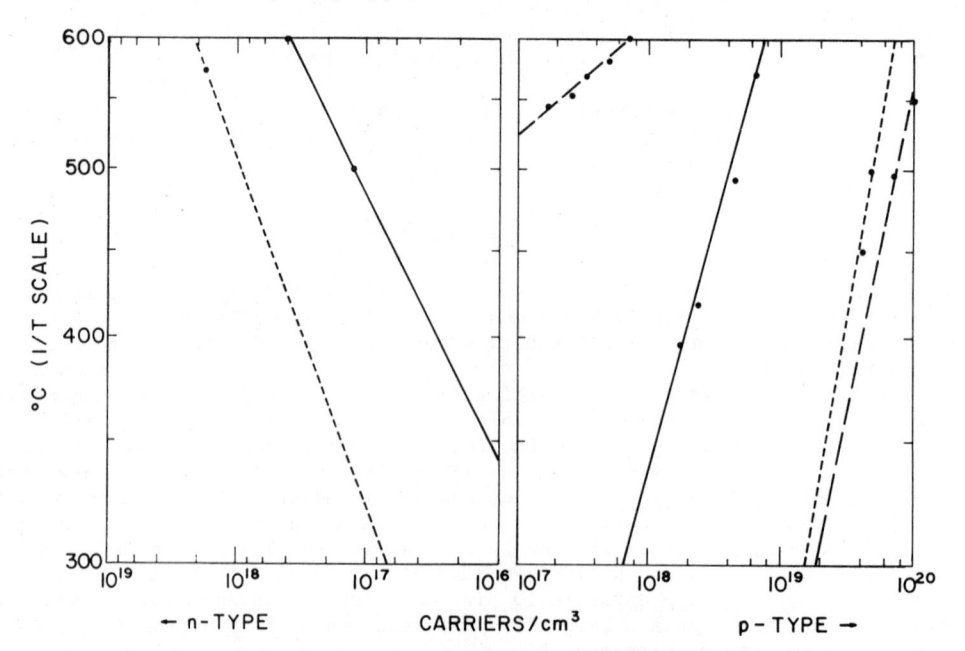

Fig. 5. Extrapolations of measured solidus boundaries for PbTe(———), $Pb_{0.8}Sn_{0.2}Te$(— —) and $Pb_{0.93}Sn_{0.07}Se$ (----); data points (·) from Ref. 26.

MBE studies show that one can indeed vary the carrier concentration over this wide range by the use of excess M or N_2 during growth (24,31,33). Figs. 6a and 6b show the relationship between excess flux and the carrier concentration of the grown film, and also compare the equivalent relationships for several impurity dopants. Data shown here are from work done in our laboratory, most of which has been reported previously (24,31). Growth on BaF_2 (34), an insulator, permitted Hall effect and resistivity measurements (at 77 K), from which carrier concentrations were calculated assuming a Hall factor of unity. Films were grown at $1.0\,mm\,hr^{-1}$ and 350°C (selenide) or 375°C (tellurides). The abscissa on Fig.6 — dopant impingement flux (atoms $cm^{-2}sec^{-1}$)/growth rate (cm sec^{-1}) — represents the dopant concentration in the grown film (atoms cm^{-3}) if all impinging dopant is incorporated. A 1 : 1 relationship between ordinate and abscissa thus corresponds to one charge carrier per impingign dopant atom, or "unity incorporation rate". The curvature at the bottom ends of all of the dopant incorporation lines on Fig. 6 indicates the effect of the baseline carrier concentration obtained for films grown from stoichiometrically-prepared binary compound evaporation sources without the use of any supplementary evaporant. These baseline levels were in the mid-$10^{16}n/cm^3$ range for PbTe and the low-$10^{17}p/cm^3$ range for the Sn-containing alloys. The slight n-type, or Pb-rich composition of the PbTe most likely results from the condensation coefficient of Te_2 being lower than that of Pb in the dissociated fraction of the impinging PbTe beam. The p-character of the Sn alloys most likely results from the non-volatility of Sn, which makes the evaporation from the SnN sources N_2-rich, as discussed in Section 1.

Looking first at the incorporation of excess N (Fig. 6a), a large excess of Te_2 is required to change PbTe to p-type, after which carrier concentration quickly saturates with increasing Te_2, at a level about an order of magnitude below the Te-rich solidus boundary predicted by Fig. 5. The low incorporation rate is expected from Fig. 2. The low saturation level may be due to annealing taking place during film cool down after growth, allowing the carrier concentration to track the solidus for awhile with accompanying precipitation of Te. The carrier concentrations of the ternary alloys increase linearly with excess Se or Te up to the high $10^{18}p/cm^3$ range, with similarly low incorporation rates as expected. Insufficient N_2 was used to observe saturation at the solidus boundaries. The impurity dopant Tl has a near-unity incorporation rate, and levels as high as 10^{20} p/cm^3 have been achieved. But in the case of PbTe, carrier concentration cannot be increased above about 10^{18} p/cm^3 without employing simultaneously impinging excess Te_2 (an atomic flux of 10% of the PbTe flux was used). Without the Te_2, carrier concentration levels out as shown and Hall mobility is also greatly reduced. These observations may be explained by the lower solubility of V_{Pb} in PbTe than in the ternary alloys (see Fig. 5). Without sufficient availability of Pb sites for Tl to enter, the remainder is forced into Te sites where it would be expected to act as a deep trap (being a group III atom on a group VI site) and thus lower the mobility. The situation is alleviated by providing a high surface concentration of excess Te with which it can react. Thus, the mode of impurity incorporation for Tl in the PbTe lattice can be controlled by N_2/M adjustment just as it can for Ge in GaAs.

Figure 6b shows a similar plot for n-dopant incorporation in the same three compounds. Given the high condensation rate of Pb predicted by Fig. 2, it is surprising that such a large excess is required to change $Pb_{0.8}Sn_{0.2}Te$ to n-type. The unincorporated excess may be observed as a precipitate on the surface even under growth conditions below the saturation level shown in Fig. 6b. The Pb does indeed, therefore, have a high condensation rate, but it also has a low incorporation rate and mostly segregates to the growth surface. The fact that $Pb_{0.8}Sn_{0.2}Te$ can be made n-type using excess Pb indicates that the M-rich solidus for this material does cross the stoichiometric line at temperatures below the data shown in Fig. 5. The impurity dopants Bi and Bi_2Te_3 are much more successful, showing near-unity incorporation rates up to 10^{19} n/cm^3 except in the case of the selenide. The nonlinear incorporation rate observed in the latter case and also in the case of Sb doping suggests the onset of compensation: a larger fraction of these group V

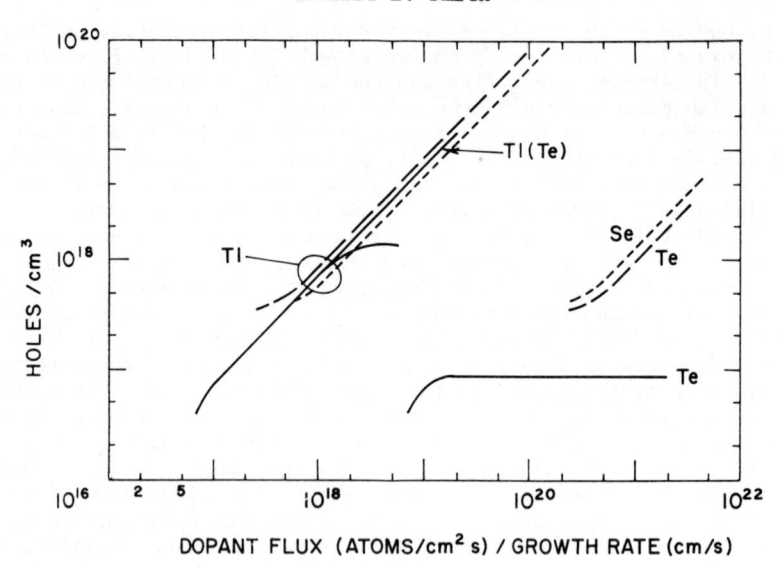

(a) p-dopants.

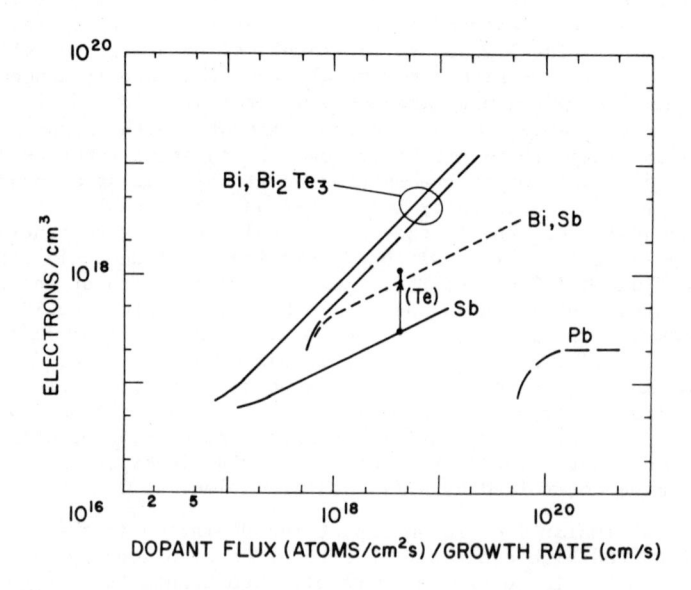

(b) n-dopants.

Fig. 6. Dopant incorporation in PbTe(———), $Pb_{0.8}Sn_{0.2}Te$ (— —) and $Pb_{0.93}Sn_{0.07}Se$(----); films grown at 1.0 μm hr^{-1}, telluride at 375°C, selenides at 350°C; (Te) indicates 10% excess Te_2 also used during impurity doping.

dopant atoms entering group VI (p-type) sites rather than group IV sites as dopant flux is increased. In one instance for Sb, 10% excess Te was found to decrease this effect, as shown. Alternatively, Bi_2Se_3 would be expected to be uncompensated if it evaporates molecularly as does Bi_2Te_3, for the Bi in this instance is already bound to Te and does not need to locate a free surface atom of Te with which to react. Were the group III-VI compounds such as Tl_2Te_3 also found to evaporate molecularly, they would be similarly useful to avoid such difficulties in p-type doping.

5. CONCLUSIONS

Native point defects caused by deviation from stoichiometry can influence the electronic properties of the compound semiconductors in a number of ways, and their concentration can be controlled over a wide range by N_2/M impingement flux ratio during MBE growth. The specific relationships between N_2/M and defect structure, including the appearance of solubility limits, are in qualitative agreement with the predictions of equilibrium thermodynamics. Control of N_2/M has been used to minimize the concentration of deep defect levels in GaAs, to regulate carrier concentration in the IV-VI compounds, and to place impurity dopants in the desired sites in both groups of compounds. Although the effects of growth temperature and growth rate on these relationships have not been studied in detail because of the relatively narrow practical ranges of adjustability of these two variables, it is likely that their fine tuning could be used to advantage to optimize defect structure while simultaneously minimizing other undesirable effects of N_2/M adjustment such as surface precipitates.

ACKNOWLEDGEMENT

The author would like to acknowledge Vincent Pickhardt for his considerable contribution to this work and Matthew Miller for many stimulating discussions, and Gerard Blom of Philips Labs for clarification of the GaAs nonstoichiometry situation.

REFERENCES

1. R.F. Brebrick, *Prog. Solid St. Chem.* <u>3</u>, 213 (1967).

2. P. Goldfinger and M. Jeunehomme, *Trans. Faraday Soc.* <u>59</u>, 2851 (1963).

3. J.R. Arthur, *J. Phys. Chem. Solids* <u>28</u>, 2257 (1967).

4. C.D. Thurmond, *J. Phys. Chem. Solids* <u>26</u>, 785 (1965).

5. R.F. Brebrick and A.J. Strauss, *J. chem. Phys.* <u>40</u>, 3230 (1964).

6. R.F. Brebrick and A.J. Strauss, *J. chem. Phys.* <u>41</u>, 197 (1964).

7. L.L. Chang and R. Ludeke, in: *Epitaxial Growth*, Part A, J.W. Matthews, (ed) pp. 37-72. Academic, N.Y., 1975.

8. R.E. Honig and D.A. Kramer, *RCA Rev.* <u>30</u>, 285 (1969).

9. D.L. Smith and V.Y. Pickhardt, *J. appl. Phys.* <u>46</u>, 2366 (1975).

10. W. Albers in: *Physics and Chemistry of II-VI Compounds*, M. Aven and J.S. Prener, (eds) Ch. 4. American Elsevier, N.Y., 1967

11. T. Yao, S. Amano, Y. Makita and S. Maekawa, *Jap. J. appl. Phys.* **15**, 1001 (1976).

12. T. Yao, Y. Miyoshi, Y. Makita and S. Maekawa, *Jap. J. appl. Phys.* **16**, 369 (1977).

13. D.G. Thomas and E.A. Sadowski, *J. Phys. Chem. Solids* **25**, 395 (1964).

14. J.R. Arthur, *J. appl. Phys.* **39**, 4032 (1968).

15. A.Y. Cho and J.R. Arthur, *Prog. Solid St. Chem.* **10**(3), 157 (Pergamon, London, 1975).

16. N. Naganuma and K. Takahashi, *Appl. Phys. Lett.* **27**, 342 (1975).

17. A.S. Jordan, A.R. Von Neida, R. Caruso, and C.K. Kim, *J. electrochem. Soc.* **121**, 153 (1974).

18. J.A. Van Vechten, *J. electrochem. Soc.* **122**, 419 and 423 (1975).

19. G.M. Blom, *J. Cryst. Growth* **36**, 125 (1976).

20. L.L. Chang, L. Esaki and R. Tsu, *Appl. Phys. Lett.* **19**, 143 (1971).

21. A.Y. Cho and I. Hayashi, *Solid St. Elec.* **14**, 125 (1971).

22. A.Y. Cho and I. Hayashi, *J. appl. Phys.* **42**, 4422 (1971).

23. A.J. Strauss, *J. Electron. Mater.* **2**, 553 (1973).

24. D.L. Smith and V.Y. Pickhardt, *J. electrochem. Soc.* (1978) **125**, 2042.

25. A.J. Strauss and R.F. Brebrick, *J. Phys. de Physique.* **29**, C4-21 (1968).

26. T.C. Harman, *J. Nonmetals* **1**, 183 (1973).

27. C.R. Hewes, M.S. Adler and S.D. Senturia, *J. appl. Phys.* **44**, 1327 (1973).

28. J.N. Walpole and R.L. Guldi, *J. Nonmetals* **1**, 227 (1973).

29. L.A. Hemstreet, *Phys. Rev. B* **12**, 1212 (1975).

30. S-S. Chern *et al.*, *J. Solid St. Chem.* **14**; 33, 44, and 299 (1975).

31. D.L. Smith and V.Y. Pickhardt, *J. Electon. Mater.* **5**, 247 (1976).

32. M.D. Miller, G.H. Olsen and M. Ettenberg, *Appl. Phys. Lett.*, **31**, 538 (1977).

33. D.K. Hohnke and S.W. Kaiser, *J. appl. Phys.*, **45**, 892 (1974).

34. H. Holloway, E.M. Logothetis and E. Wilkes, *J. appl. Phys.* **41**, 3543 (1970).

THE AUTHOR

Dr. Donald L. Smith

Dr. Smith's formal training is in Chemical Engineering: B.S., M.I.T., 1965; Ph.D., University of California at Berkeley, 1969. At Berkeley he studied the catalytic behavior of single-crystal platinum using molecular beams, mass spectroscopy and LEED. For the past 8 years he has been on the research staff at Perkin-Elmer and has been mostly involved with the growth of various dielectric, semiconductor, and metallic thin films. His particular interest has been the MBE growth of the II-VI, IV-VI, and III-V semiconductors for various electro-optic device applications, and he has presented several invited papers on the subject in addition to his contributed articles. He makes his home in Westport, Connecticut, with his wife Jane Mickelson, who is an author, reviewer, and lecturer on contemporary fiction, and with numerous children and one cat.

MBE TECHNIQUES FOR IV-VI OPTOELECTRONIC DEVICES*

H. Holloway

Ford Motor Company, Dearborn, Michigan 48121, U.S.A.

and

J.N. Walpole

Lincoln Laboratory, Massachusetts Institute of Technology, Lexington, Massachusetts 02173, U.S.A.

(Submitted July 1978)

ABSTRACT

Recent development of high-quality IV-VI optoelectronic devices grown by MBE has significantly increased the technological importance of this epitaxial technique in IV-VI materials. Despite the progress made, much is still unknown about the importance of crystal perfection for device performance. Criteria for crystal perfection required to minimize carrier recombination and optical losses in i.r. devices in IV-VI materials need to be established.

In this paper the literature on IV-VI film growth by vacuum deposition techniques is briefly reviewed. MBE techniques used for growth of IV-VI materials on BaF_2 and SrF_2 substrates and for growth of PbSnTe on PbTe substrates are described. Emphasis is on the techniques used to deposit pseudobinary alloys with homogeneous composition. The use of foreign impurity dopants is also discussed. Criteria of quality and crystal perfection of epitaxial layers needed for device performance are evaluated. In layers grown on IV-VI substrates, large dislocation densities due to lattice mismatch are normally present but may not be detrimental to device performance. Lattice-matched heteroepitaxial systems are discussed. For insulating substrates, it is shown that the crystalline quality and the carrier mobility of layers grown on BaF_2 and SrF_2 are superior to those grown on alkali halide substrates and that the latter are unsuitable for making devices.

Carrier mobility is not a good test of crystalline perfection, however, it is argued that device performance, particularly in the demanding requirements for low-noise photodiodes, is a sensitive measure of the epitaxial material quality. Thin-film photodiodes grown by MBE and the hot-wall techniques are then reviewed in detail, including some recent unconventional

---

* The Lincoln Laboratory portion of this work was sponsored by the Department of the Air Force and the Defense Advanced Research Projects Agency.

devices in which the unique properties of thin films are
exploited. Heterostructure optical waveguides are described
briefly followed by a review of diode laser results obtained
by MBE and the hot-wall technique. These results are for
thin-film PbTe lasers grown on BaF_2 and for a variety of
devices grown on I-VI substreates including single heterostructure, double heterostructure, homostructure, and
distributed feedback lasers.

1. INTRODUCTION

The first major application of the IV-VI semiconductors was the use of PbS layers
as infrared-sensitive photoconductors in the 1940's. Most of the IV-VI photoconductive work has been with polycrystalline precipitates, which represent a diver-
gent evolutionary line from the optoelectronic devices that are described in the
present article. However, the early studies of photoconductive PbS included vacuum
deposition of thin films (1,2), which led to a demonstration of epitaxial growth on
NaCl (3) that may be regarded as the progeniotor of modern epitaxial IV-VI devices.
In a second wave of development during the 1960's, techniques were established for
the growth of goodquality single crystals of IV-VI semiconductors and for making
p-n junction devices that are discussed here.

The selection of subject matter for the present review involves some arbitrarily
chosen boundaries because there is not a well-defined separation between thin-film
and bulk crystal growth techniques. At one end of the spectrum we have vacuum
deposition of μm thick films using a molecular beam that is generated with an effusion
cell. At the other extreme, we find closed-tube transport down a temperature gradient
to give crystals with cm dimensions either by spontaneous nucleation or with use of
a seed crystal (4). This range of techniques includes both epitaxy on isostructural
IV-VI substrates (or seeds) and or insulating, such as BaF_2.* Much of the recent
literature on growth of IV-VI thin films relates to a tehnique in which the source
and the substrate are enclosed by a "hot-wall" that pipes most of the evaporant to
the substrates. ** This method may be regarded as intermediate between molecular-
beam epitaxy (MBE) *** and closed-tube transport. Thus, depending upon the wall
temperature, the efficieny of the piping and the approximation to equilibrium, growth
conditions may vary between the extremes that are obtained with MBE and with closed-
tube transport. This intermediate character is also evident in the hot-wall layer
thicknesses which may exceed 100 μm (8), thereby approaching bulk crystal growth.

In the following, the description of epitaxial growth techniques is restricted to two
representative MBE systems that were developed by the authors and their colleagues
to pursue their respective interests in epitaxial i.r. detectors and epitaxial i.r.
lasers. We have mostly omitted reference to extensive body of earlier work on MBE
of IV-VI semiconductors because, despite its value for fundamental studies of the
materials, the emphasis on alkali halid substrates led to layers whose crystal
perfection was inadequate for p-n juncation devices.

---

*   Insulating substrates are usually associated with tin films of the IV-VI
semiconductors, but Pandy[5] has shown that closed-tube growth of (Pb,Sn) Te may
be initiated with a BaF_2 seed.

**   The hot-wall technique appears to have originated with Koller and Gohill (6) who
used it to deposit ZnS. Subsequently it has been used extensively for IV-VI
semiconductors (7-14).

*** Molecular-beam epitaxy is a recent designation for a much older technique. This name is quite descriptive for the IV-VI semiconductors where the beams are really of molecular species. The term is also applied (15), though with less justification, to growth of materials like the III-V semiconductors, where one works with beams of the constituent elements.

These studies have been reviewed by Zemel (15b). Epitaxial growth by the hot-wall technique has also been omitted because a recent review is available (16). However, for completeness we have included the results of recent hot-wall studies which indicate that this technique can yield devices whose performance is comparable to those made with MBE. We have excluded consideration of liquid-phase epitaxy because this would take us too far from the main area of our review, but we note in passing that this method has been shown to give high-performance IV-VI photodiodes (17,18) and lasers (19) on IV-VI substrates.

2. TECHNIQUES FOR MOLECULAR-BEAM EPITAXY

Vacuum deposition of the IV-VI semiconductors and their pseudobinary alloys has mostly been accomplished by evaporative methods, although some use has been made of flash evaporation (20) and of sputtering (21). From the literature one may conclude that almost any of the standard vacuum deposition techniques will yield epitaxial layers, so that the choice of method is to some extent a matter of taste. However, evaporative techniques are particularly attractive because of the IV-VI semiconductors have convenient vapor pressures ($P \approx 0.1$ Torr at 1000 K) and sublime predominantly as diatomic molecules (22-24).

Even with evaporative techniques, care is needed when the deposit is a pseudobinary alloy. Use of a source of the alloy (25,26) does not, in general, give truly congruent evaporation. For example, Northrop (27) has shown that the gas phase in equilibrium with $Pb_{1-x}Sn_xTe$ is enriched in the SnTe component. The enrichment is roughly 30% over the range of compositions ($x < 0.2$) that are suitable for i.r. detectors. With use, a $Pb_{1-x}Sn_xTe$ source that is operated near equilibrium will become depleted of SnTe and the energy gap of the deposit will drift towards larger values. This places a severe constraint upon the fraction of the source material that may be used if a uniform deposit composition is required. From Northrop's results a source with the composition $Pb_{0.85}Sn_{0.15}Te$ when operated near 700°C will initially give a deposit with composition near $Pb_{0.80}Sn_{0.20}Te$ corresponding to a cut off wavelength of approximately 11 μm at 80 K. After use of 20% of the source material the source and deposit compositions may be estimated to be $Pb_{0.86}Sn_{0.14}Te$ and $Pb_{0.81}Sn_{0.19}Te$, respectively.. This would lead to a decrease of the 80 K cutoff wavelength of the depositied semiconductor by approximately 0.5μm.

Evaporation of a pseudobinary alloy without a drift in its composition is possible, provided an open crucible rather than a Knudsen cell us used, and provided the source material is made homogeneous and is in a solid (rather than granular or powered) form. Diffusion of the alloy composition can then become sufficiently small compared to the rate of removal of the surface material that the solid never reaches equilibrium with the vapor. Electron microprobe analysis (28) of an angle-lapped (5°) 8-μm film grown on a PbTe substrate at approximately 380°C from a $Pb_{0.88}Sn_{0.12}Te$ alloy source showed a constant composition throughout the film of 0.120 ± 0.001 SnTe.* At the interface of the film with the substrate, the profile shown in Fig. 1 was obtained. The errors in SnTe composition are based on the statistical scatter obtained in four different traces through the region.

---

* Strauss (29) has reported larger variations in film composition in growth of PbSnSe films from pseudobinary alloys. However, the source ingots were not subjected to a homogenizing high-temperature anneal.

The error in distance of ± 0.5 μm is an estimate of the minimum error due to penetration of the microprobe into the angle-lapped surface. Larger errors are likely. The lapping produces some smearing of the probed surface which may lead to additional errors. In any case, the transition region appears to be < 1μm. Since this film was grown at 1μm hr^{-y}, the 8-hr growth time represents an unusually long period and the diffusion of the Pb-Sn composition should be the maximum normally encountered.*

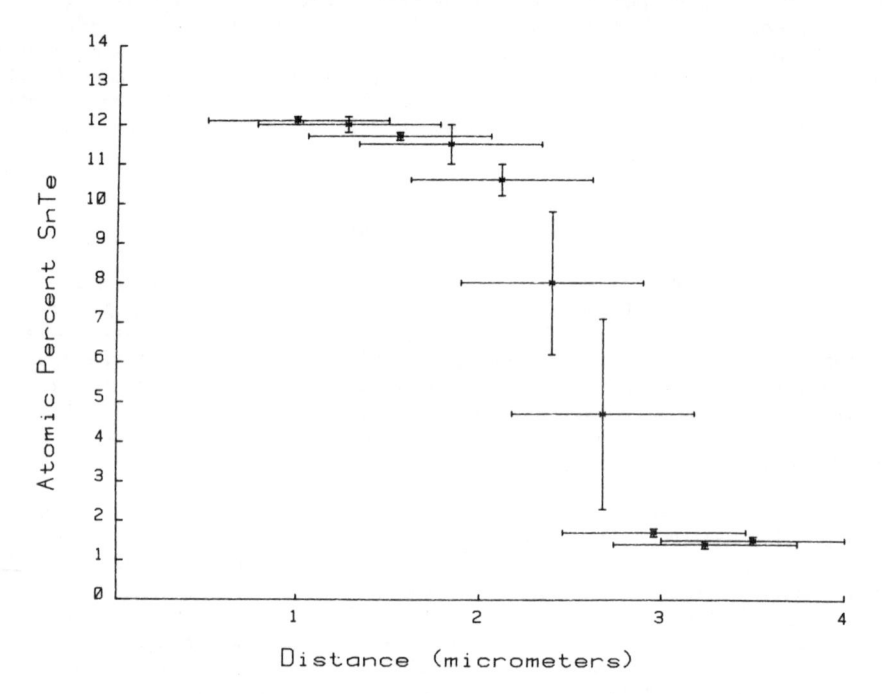

Fig. 1. Variation of alloy composition with distance near the interface of a $Pb_{0.88}Sn_{0.12}Te$ layer on a PbTe substrate as determined by electron microprobe analysis on an angle-lapped (5°) surface. The uncertainty in distance is the minimum estimated error. The absolute SnTe composition by ≈2% although relative changes can be more accurately determined as indicated by the vertical error bars.

The vacuum system used for MBE of PbSnTe on PbTe substrates is shown schematically in Fig. 2. Ion pumping with a system pressure during growth of about 10^{-7} Torr was used. The substrate was radiently heated in a small furnace open at one end. The rate of deposition was controlled by a feedback signal to the source-heater power supply. The feedback signal was generated from the deposition rate onto a quartz-crystal film-thickness monitor. Rates of 1-3 μm hr^{-1} were used with substrate temperatures of 350 - 450°C. The sources were contained in open pyrolytic boron nitride crucibles heated in a tantalum furnace. The distance between the sources and the substrate was about 20 cm. Separate sources were used for each type of layer grown. Each source was prepared form zone-melted (32) binary compounds

---

* Bicknell (30,31) has used X-ray techniques to investigate grading in PbSnTe films grown by a vapor epitaxy technique at higher substrate temperature (500-600°C). He finds somewhat more Pb-Sn composition diffusion.

(PbTe and SnTe) weighed out in the proportions required to obtain the desired alloy composition.

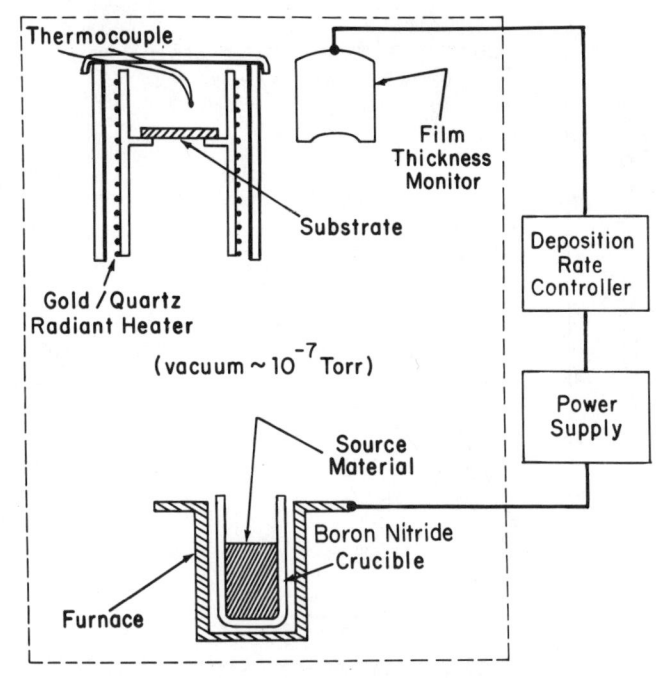

Fig. 2. Schematic diagram of vacuum deposition system for epitaxial growth from PbSnTe alloy sources on PbTe substrates.

Before use, the sources were homogenized by sealing the constituents in an evacuated quartz ampoule and heating above the melting point for several hours followed by rapid quenching in water to room temperature. Subsequently, the ingot was annealed for 5 days below the melting point at 700°C and quenched again. Pieces of the polycrystalline ingot were then broken off and used as solid chunks in the source furnaces.

When undoped sources were used, very low concentration films were obtained as determined by waveguiding experiments discussed below. Foreign impurities were used to control carrier type and concentration rather than controlling the deviation from stoichiometry. Impurities are expected to diffuse much less than the deviation from stoichiometry (33).

For n-type films, Bi was added in controlled amounts and for p-type films TlSe was added. Figure 3 shows the carrier concentration in the film compared to the concentration of dopant in the source. Above about 5×10^{18} cm^{-3} the film concentration appears to saturate for both types of dopants. The film concentration was measured using the carrier concentration dependence (34) of the infrared reflectivity minimum at the plasma edge.

The use of separate effusion sources for the components of pseudobinary alloys* may also be troublesome because the ratio of the component fluxes must be precisely controlled. If one assumes an effusion cell that operates at about 1000 K, a typical temperature fluctuation of ± 1 K together with a sublimation enthapy of about 50 kcal mole^{-1} gives a fluctuation $\pm 2.4\%$ in the effusion rate. With two independent sources the overall fluctuation in the ratio of the components is then $\pm 5\%$. This is unacceptable in cases where the energy gap is a sensitive function of composition.

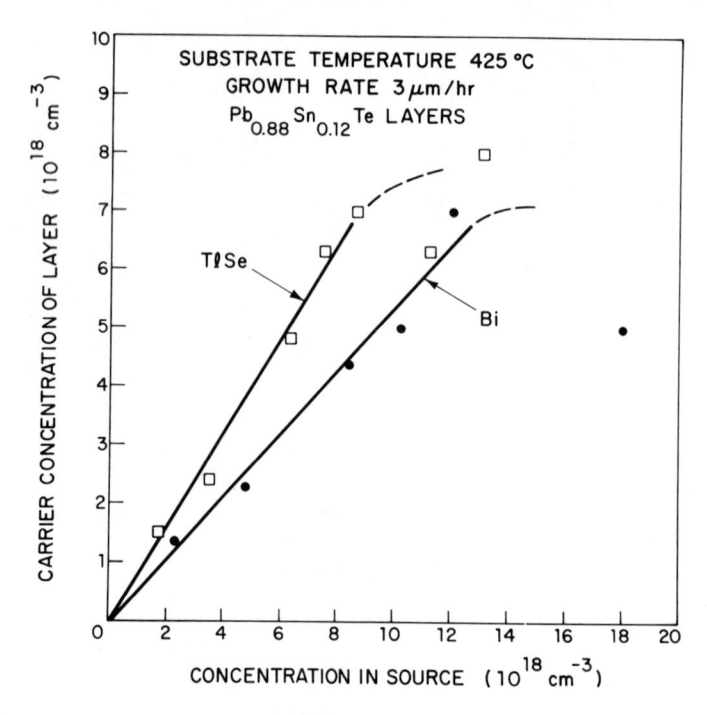

Fig. 3. Calibration curves of Bi and TlSe doping efficiency in $Pb_{0.88}Sn_{0.12}Te$ films.

Composition fluctuations may be reduced by using a pair of thermally-linked cells (36). The idea is, that once the temperature fluctuations of the two cells have been minimized, their effect upon the composition can be further reduced by having them occur in phase. A change in the effusion rate of one of the components is then partially offset by a corresponding change in the effusion rate of the other component. For small temperature fluctuations the fractional change in the effusion rate of one of the components may be expressed as

$$\frac{\Delta J}{J} = \frac{1}{J} \frac{\partial J}{\partial T} \Delta T \underset{\sim}{\sim} \frac{\Delta H_s}{RT^2} \Delta T,$$

---

* It is also possible to use separate effusion sources for the elemental constituents (35), but this seems unnecessarily complicated because, apart from the increase in the number of cources, the large range of volatility of the constituents makes radiation shielding between the sources necessary. (E.g. for growth of (Pb,Sn)Te the equilibrium vapor pressures of the constituent elements at 1000-K are: Pb, 10^{-2} Torr; Sn, 5×10^{-8} Torr; and Te, 2 Torr. In contrast, the binary compounds PbTe and SnTe both have vapor pressures of about 0.1 Torr at 1000-K).

where ΔH_s is the enthalpy of sublimation. For a pair of effusion cells that operate at the same temperature, but with independent random temperature fluctuations, the maximum change in the flux ratio is:

$$\Delta \left.\frac{J_1}{J_2}\right|_{max} \approx \pm \frac{J_1}{J_2} \frac{(\Delta H_{s,1} + \Delta H_{s,2})}{RT^2} \Delta T,$$

where $\Delta H_{s,1}$ and $\Delta H_{s,2}$ are the enthalpies of sublimation of the two components. If the temperature fluctuations are now constrained to occur in phase, the maximum change in the flux ratio is reduced to:

$$\Delta \left.\frac{J_1}{J_2}\right|_{max} \approx \pm \frac{J_1}{J_2} \frac{(\Delta H_{s,1} - \Delta H_{s,2})}{RT^2} \Delta T.$$

The magnitude fo the effect may be estimated by again considering the deposition of $Pb_{1-x}Sn_xTe$ with $E_g \approx 0.11$ eV. The enthalpies of sublimation of PbTe and SnTe are 52.3 kcal mole^{-1}, respectively. With a pair of effusion sources that operate at 1000 K with independent random fluctuations of 1 K, the flux ratio can change by $\pm$ 5% giving $\Delta x = \pm$ 0.01 and $E_g = \pm$ 6 meV. In contrast with in-phase temperature fluctuations, the flux ratio changes by about* $\pm$ 0.3% to give $\Delta x \approx \pm$ 6 X 10^{-4} and $\Delta E_g \approx \pm$ 0.4 meV. In practice, the temperature control can be improved to about $\pm$ 0.2 K with a corresponding further improvement in homogeneity. Schmatics of the deposition system and the double cell are shown in Figs. 4 and 5.

3. THE CRYSTAL PERFECTION OF IV-VI EPITAXIAL LAYERS

The work described in this review has made use of IV-VI semiconductor substrates, principally for work with lasers, and insulating substrates, principally for work with photodiodes. There is relatively little prior information on the properties of the layers on IV-VI substrates, largely because good single crystal substrates are not readily available.**

With IV-VI substrates, it is possible in principle to achieve a match in the lattice constants of the substrates and epitaxial film. This may be technologically important as a means of reducing the density of the misfit dislocations that occur in such IV-VI structures (37,39). Figure 6 shows the lattice constant vs energy gap for the lead-tin chalcogenide compounds and alloys of most interest. The lines representing the ternary alloys have been drawn as straight lines between the points representing the binary compounds and hence are only approximate. In a study of heteroepitaxy of $Pb_{1-x}Sn_xTe$ on {100} substrates of different alloy composition using both MBE and LPE (40) it was found that etchpit densities in layers several μm in thickness were invariably greater than 10^7 cm^{-2} even for differences in tin composition as small as 2%. Homoepitaxy or nearly latticematched heteroepitaxy of

---

* There is a significant uncertainty here because the possible errors in the sublimation enthalpies are comparable to the 6 kcal value for their difference.

** The unsuitability of these conducting substrates for the transport measurements which have dominated such work may also have had an influence.

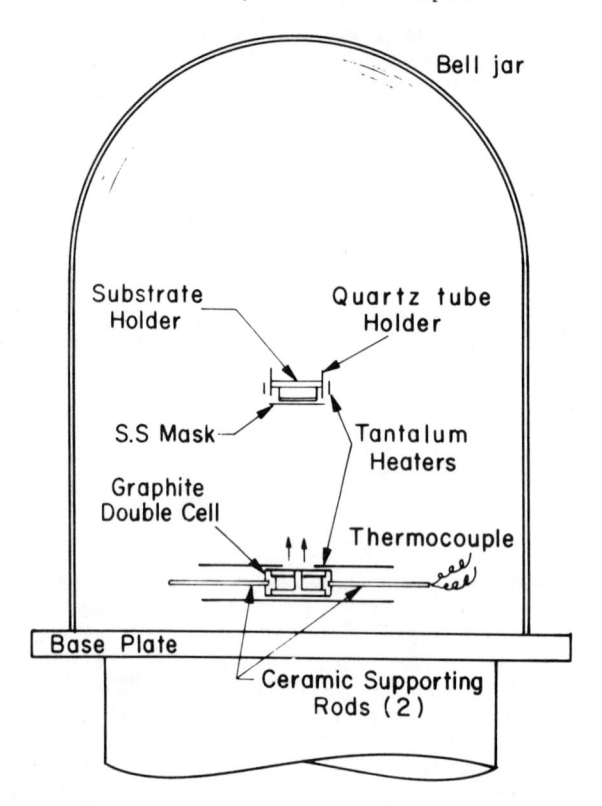

Fig. 4. Schematic diagram of a vacuum deposition system for epitaxial growth of IV-VI semiconductors on fluorite structured substrates.

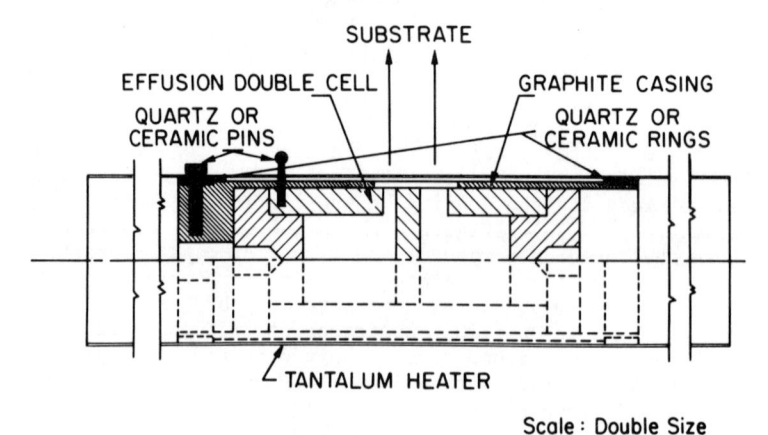

Fig. 5. Isothermal double cell for evaporation of IV-VI pseudobinary alloys.

$Pb_{0.88}Sn_{0.12}Te$ on $PbTe_{0.952}Se_{0.048}$ (or vice versa) yielded etch-it densities comparable to that of the substrates used, $< 10^4$ cm^{-2}.

Lattice-matched heteroepitaxy of PbSnSe on PbSSe has not been reported but appears particularly attractive since both these alloys have good metallurgical and mechanical properties especially compared to PbTeSe. To date, no optoelectronic devices have been reported which were grown using lattice-matched heteroepitaxy. However, the device results to be discussed suggest that misfit dislocations may have little effect on minority carrier liftime in films* and there is evidence that recombination at hetero-interfaces can be quite small.

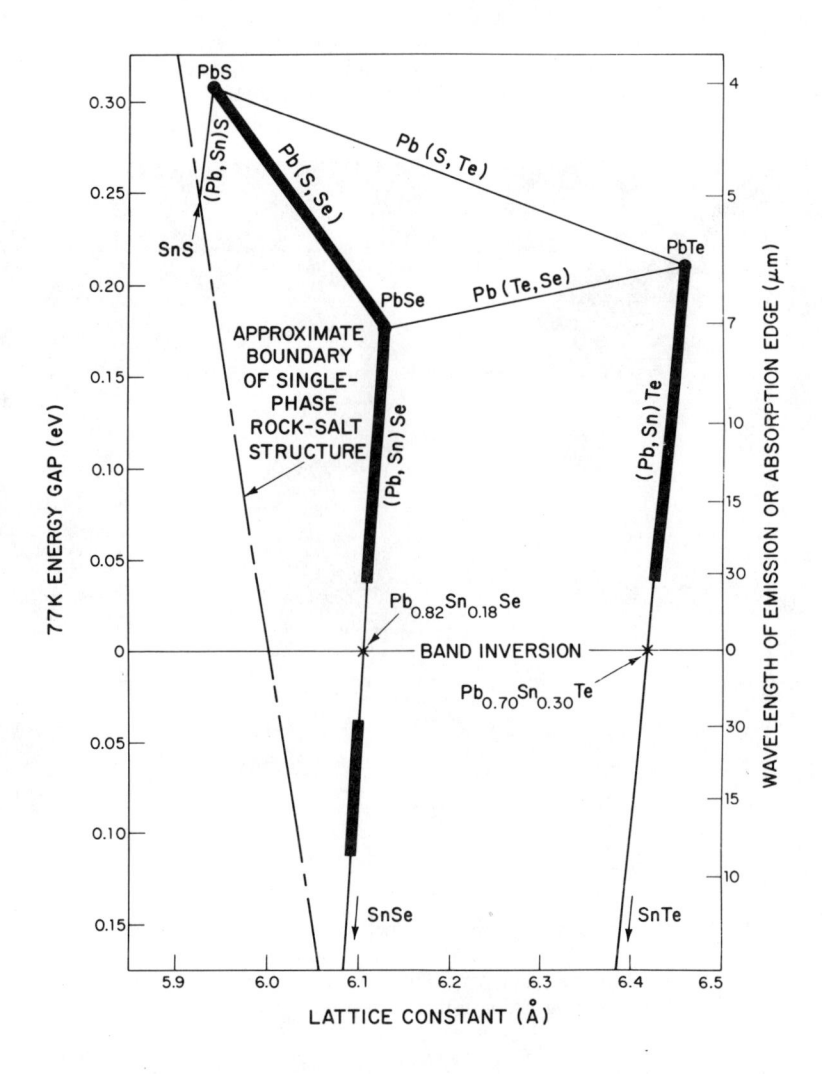

Fig. 6. Variation of energy gap at 77 K (wavelength of emmission or absorption edge) with latice constant for the lead-tin calcogenide alloys.

---

* Bicknell (41) has shown that misfit dislocations in vapor epitaxial layers in PbSnTe have no effect on minority carrier dissuion length.

In contrast to the situation with IV-VI substrates, the growth of IV-VI layers on insulating substrates has a substantial literature (15) that is mostly devoted to growth on alkali halide substrates, although occasional use has been made of other materials, such as CaF_2 and mica. This application of such semiconducting thin films to p-n junction device technology demands a more critical approach to crystalline perfection than has been customary in the literature on epitaxial growth. Determination of the degree of epitaxy in IV-VI layers has been mostly by diffraction techniques that are insensitive to the spread in orientation that is associated with a small-grained mosaic structure. An early study (42) of PbS on NaCl demonstrated the presence of a mosaic structure with grain sizes of the order of 1000 . Similar results were obtained later with PbTe on NaCl (43) which was found to have about 5×10^{10} dislocation/ cm^2 that were mostly arranged to form grain boundaries with about 2000 spacing and about 2° misorientation. An electron micrograph that reveals the typical grainy structure of a PbTe layer on an alkali halide substrate is shown in Fig. 7.

Fig. 7. Parlodion replica of epitaxial PbTe grown on cleaved KCl. The region shown is about 40 μm wide and the darker regions are PbTe grains that have been pulled out of the film during replication.

Despite the early indications of a mosaic structure in epitaxial IV-VI semiconductors on alkali halide substrates, much of the subsequent work ignored the crystal perfection of the layers and, consequently, has little relevance to p-n junction device applications. The situation resembles that in the early stages of development of epitaxial III-V semiconductors, such as GaAs, and earlier somments (44) about the inadequacy of conventional diffration techniques for assessment of the crystal perfection of II-V layers apply with equal force to the IV-VI semiconductors. Such problems of characterization appear to be widespread with epitaxial semiconductors.

A commonly reported property of IV-VI epitaxial layers is the Hall mobility. Layers that have been prepared under a wide range of conditions on various substrates tend to exhibit bulklike values of the Hall mobility at 300 K and, to a lesser extent, at temperatures down to 77 K. However the scattering lengths in these materials are too small for the carrier mobilities to give significant information about the crystal perfection. In the temperature range 300-77 K, the largest mobilities that are observed with the lead chalcogenides follow,

$$\mu_H \propto T^{-a}, \quad a \approx 5/2,$$

and this relationship has been interpreted in terms of scattering by phonons (45). At lower temperatures the mobility tends to saturate at values that have been attributed to scattering by ionized native defects (46).

For the lead chalcogenides, the phono-limited mobilities of both p-and n-type materials are about 2.5×10^4 cm^{2}v^{-1}sec^{-1} at 77 K. For typical carrier concentrations in the range $10^{16} - 10^{18}$ cm^{-3} this corresponds to scattering lengths of only about 2000 . Thus, the 77 K Hall mobility cannot give information about departures form perfect crystallinity on a scale greater than a few thousand . Even at 4 K, where the Hall mobility of PbTe may be of the order of 10^6 cm^{2}v^{-1}sec^{-1}, the scattering lengths still do not exceed a few μm. The utility of the mobility as a criterion for the perfection of epitaxial layers is further reduced by the fact that even good device-quality bulk crystals frequently fail to attain the full phonon-limited mobility at 77 K. Thus, high performance bulk crystal PbTe photodiodes have been made from p-type material with $\mu_H(77 K) = 1.0 \times 10^4$ cm^{2}v^{-1}sec^{-1} (47) rather than the phonon limited value. These departures form the phonon-limited mobility appear to correlate with the thermal history of the specimen and it has been suggested that they arise from varying degrees of ionized impurtiy scattering by complimentary pairs of native defects (46). Similar effects have been observed with epitaxial PbTe and Pb$_{0.8}$Sn$_{0.2}$Te.

In the evaluation of IV-VI semiconductors on insulating substrates as materials for p-n junction devices it is important to verify that the layers are truly single-crystalline, rather than the fine-grained mosaics with some degree of epitaxial alignment that are commonly designated "single-crystal films". Beyond this point, characterization is made difficult by a lack of quantitative relationships between imperfections, such as low-angle grainboundaries, and the properties of p-n junctions, which are more easily related to such parmeters as carrier lifetime and diffusion length. Thus, at present the ultimate test of an epitaxial layer is provided by comparing the properties of a p-n junction in it with those of a similar junction in a bulk crystal of the same semiconductor.

Epitaxial growth of Pb$_{0.8}$Sn$_{0.2}$Te on BaF$_2$ was first reported in 1970 (48). Subsequent work (49) showed that thse and similar layers of PbTe were remarkably free of low-angle grain boundaries. Figure 8 shows an electron micrograph of a replica of a PbTe layer grown on cleaved BaF$_2$. The most evident features are triangular pits that appear to arise at the intersection of dislocations with the layer surface.*
The dislocation densities so estimated can vary quite widely from nearly 10^8 cm^{-2} to significantly less than 10^6 cm^{-2}. Occasional low angle grain boundaries, as shown in Fig. 8, appear to arise by propagation of boundaries in the substrate. The typical misorientations across these isolated boundaries, as estimated from the dislocation spacing, are a few minutes arc and correspond with those found in some

---

* The dislocation pits may not occur if the substrate temperature is too low of the growth rate is too large. We have observed well-developed pits with growth at 1-2 μm hr^{-1} with substrate temperatures in the range 300-400°C.

of the substrates.

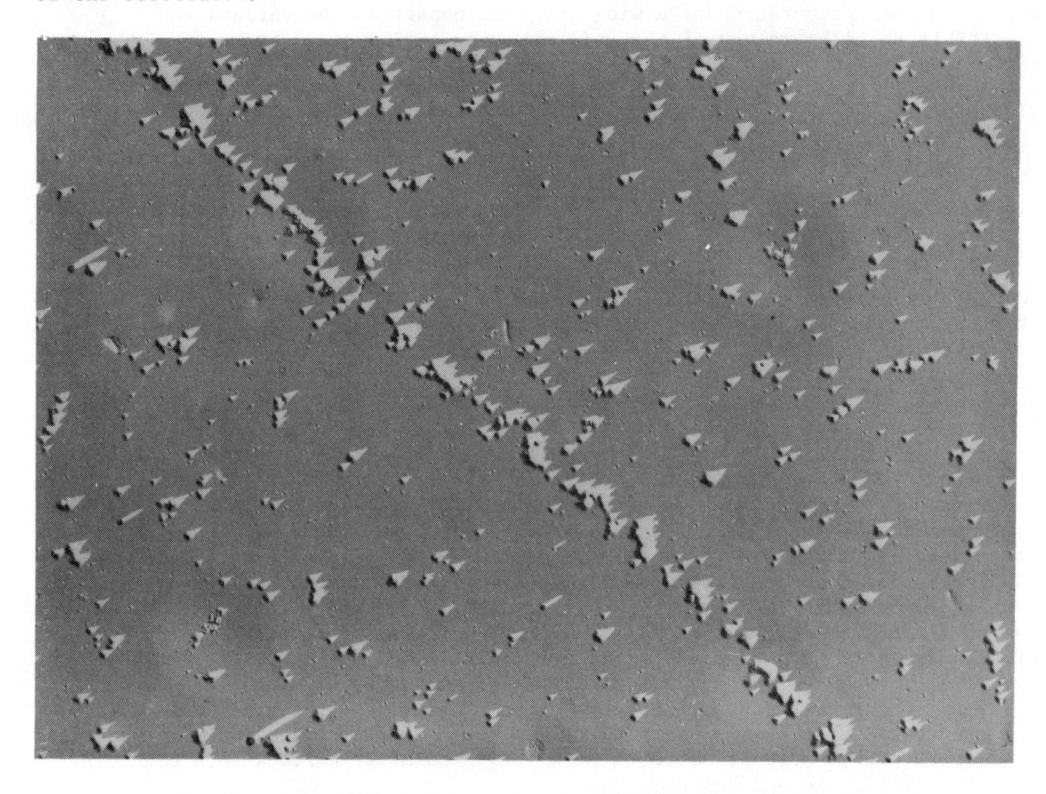

Fig. 8. Parlodion replica of epitaxial PbTe grown on cleaved BaF_2. The region shown is about 40 μm wide and the line of dislocation pits corresponds to a tilt of a few minutes arc.

Table 1 gives a comparison of some properties of the fluorite structured and the rock-salt structured substrates. Adequate lattice matches with the lead chalcogenides are obtained with both substrate types. The {111} cleavage of the fluorite structures leads to an unfamiliar growth habit for the rock-salt structured semiconductors, but this appears not to give problems provided that the vacuum system is sufficiently clean. (Insufficient care sometimes leads to the more familiar {100} habit as a fibre texture). The fluorite structured substrates do give a superior thermal expansion match and this initially prompted the trial of BaF_2 substrates in an attempt to reduce damage when the layers were taken from a growth temperature near 700 K to an operating temperature near 80 K. However, the thermal expansion mismatch does not appear to provide an explanation for the relatively grainy films that are obtained with alkali halide substrates. A significant feature may be the hydroscopic nature of the alkali halide surfaces. In contrast, cleavage surfaces of the fluorite homologues are stable in moist air and their low volatilities permit vacuum bake-out at relatively high temperatures (up to at leasts 800 K) to allow clean-up of air-cleaved surfaces.

An unusual feature of the IV-VI layers on fluorite structured substates * is that epitaxy occurs with the substrate and deposit in a twinned rather than a parallel relationship, i.e. growth on the {111} BaF_2 surface gives a PbTe deposit that is rotated 180° about the {111} axis relative to the substrate (51,52). The result may be interpreted in terms of an interface structure of the type

$$A\ \beta\ \alpha\ B\ \gamma\quad \beta\ C\ \alpha\ \gamma\ A\ \beta\quad C\ \alpha\ B\ \gamma\ A\ \beta\ C$$
$$\underline{\hspace{2cm}BaF_2\hspace{2cm}}\quad\underline{\hspace{1.5cm}PbTe\hspace{1.5cm}}$$

where the stacking symbols (A,B,C for metals; α,β,γ for non-metals) have their usual significance. A detailed analysis shows that alternative interfacial structures are less favoured both electrostatically and sterically (53).

In general, Hall measurements of IV-VI epitaxial films on fluorite structured substrates give results that differ little from those reported for the best bulk single crystals. As discussed above, the attainment of large Hall mobilities is not an adequate criterion for the growth fo device-quality material. However, it is of interest that the low-temperature mobilites obtained with BaF_2 and SrF_2 substrates are significantly larger than those that have been reported for layers on alkali halide substrates. A typical example is the result for PbS (54) that is shown in Fig. 9. For both n- and p-type layers the films give temperature-dependent mobilities that agree well with the best results obtained with bulk PbS. Surpirsingly, the mobility of the n-type layers is not significantly affected by growth from Pb-rich fluxes that appear to give Pb precipitates in the layers. A similar insensitivity to precipitates has been observed with n-type PbSe layers that were grown without a subsidiary Se source.

Table 1. Properties of the lead chalcogenides and some substrates

Compound	Lattice Constant ( )	α ($K^{-1} \times 10^6$) near 300 K	Cleavage	P(Torr) at 700 K
PbS	5.94	20		
PbSe	6.12	19		
PbTe	6.46	20		
NaCl	5.64	39	{100}	4.5×10^{-7}
NaBr	5.96	42	{100}	3.7×10^{-6}
NaI	6.46	45	{100}	4.0×10^{-5}
KCl	6.29	37	{100}	2.2×10^{-6}
KBr	6.59	38	{100}	1.1×10^{-5}
KI	7.05	40	{100}	3.9×10^{-5}
CaF_2	5.40	19	{111}	$\sim 6 \times 10^{-20}$
SrF_2	5.80	18	{111}	$\sim 1 \times 10^{-20}$
BaF_2	6.20	18	{111}	$\sim 3 \times 10^{-17}$

* This applies to BaF_2 and SrF_2. While lattice match is not particularly critical, it appears that CaF_2 has too large a mismatch for successful epitaxy of the IV-VI semiconductors. Studies in the Ford Laboratory have failed to give useful layers on CaF_2 substrates and this is in accord with published results for (Pb,Sn)Te on CaF_2 (50).

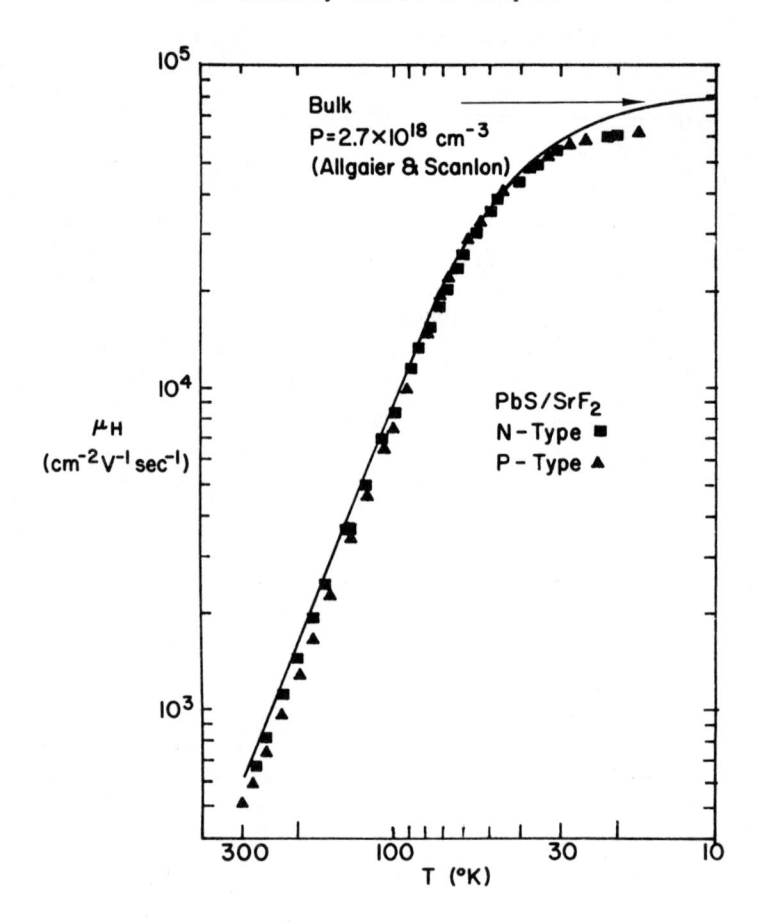

Fig. 9. Hall mobilities of PbS layers grown on cleaved SrF$_2$ substrates. The curve shows the largest mobility that has been reported for a bulk crystal of PbS (Ref. 22).

With PbTe and (Pb,Sn)Te layers on BaF$_2$, increases in the mobility have been obtained by annealing the films at relatively low temperatures (typically 300-400°C for 12-15 hr) (49). The increase was in the more-or-less temperature-independent saturation mobility that was attained at lower temperatures, rather than in the phonon-limited mobility, for which $\mu_H \propto T^{-\alpha}$ with $\alpha \approx 5/2$. The annealing results were interpreted in terms of a reduction in the ionized defect scattering that accompanies the pairwise recombination of native donors and acceptors. It was postulated that the layers as grown are compensated semiconductors with about 10^{18} cm^{-3} of native defect pairs although their carrier concentrations are typically of the order of 10^{17} cm^{-3}. More recent results suggest that such a level of compensation is not invariably present in the layers. Thus, Fig.10 shows Hall data (54) for an as-grown 2.7 μm thick n-type PbTe layer on BaF$_2$ that attains a low-temperature mobility of 5 x 10^5 cm^{2}v^{-1}sec^{-1}. This appears to be the largest mobility that has been reliably established for a IV-VI epitaxial layer.*

+ Lopez-Otero (55) has reported low-temperature Hall mobilities of up to 2.5 x 10^6 cm^{2}v^{-1}sec^{-1} in PbTe layers grown on BaF$_2$ and suggests that this result reflects the superiority of the hot-wall method over other deposition techniques. However, these results must be regarded as questionable because the large mobilities at

13 K were obtained with specimens whose (56) mobilites at 77 K exceeded the commonly observed phonon-limited value by about a factor of two. Similar results with bulk single crystals of (Pb,Sn)Te (57) have been interpreted as a consequence of inhomogeneity in the specimens (58).

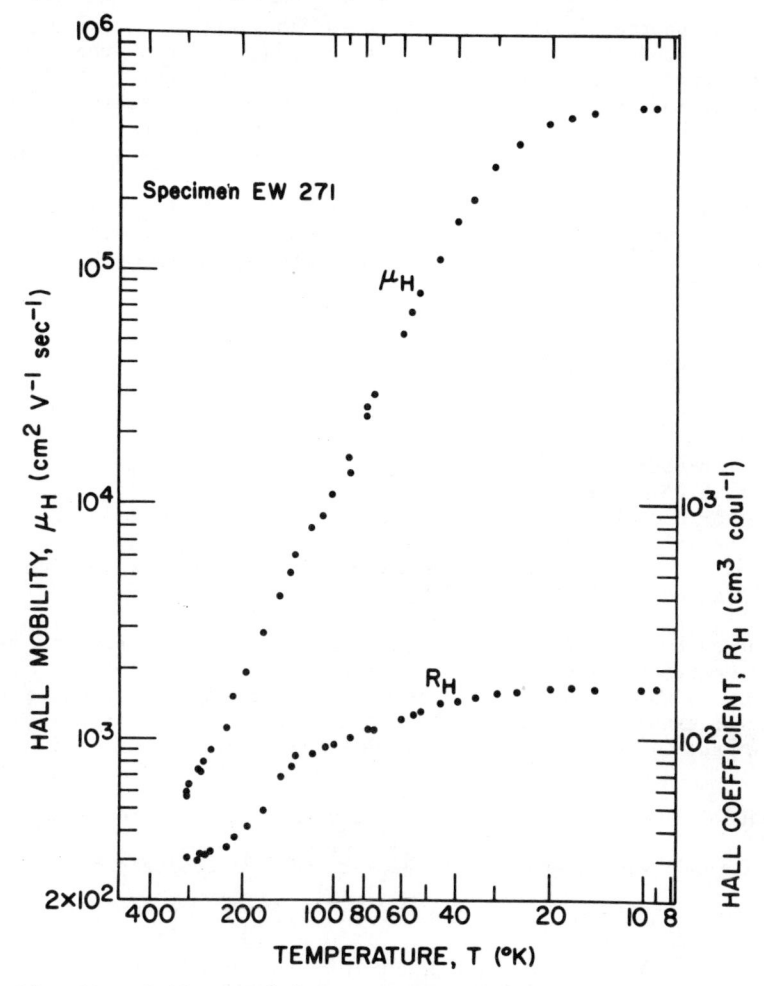

Fig. 10. Hall coefficient and Hall mobility of a 2.7 μm-thick n-type PbTe layer on a BaF$_2$ substrate.

4. PHOTODIODE PERFORMANCE AS A CRITERION FOR CRYSTAL QUALITY

As pointed out earlier, the ultimate test of a semiconducting material arises when one attempts to make devices with it. For discussion of the results that have been obtained with thin-film photodiodes it is convenient to consider the detectivity (D^*), which is a standard figure of merit that is defined as a signal-to-noise ratio that has been noralized to make it independent of the incident power, the noise bandwidth, and the detector area.

With, a well-behaved photodiode at zero bias in the absence of significant blackbody radiation form the background, the signal current depends upon the quantum efficiency (η) and the noise current depends upon the Johnson noise of the junction resistance. Under these conditions we have

$$D^*(\text{Johnson}) = \eta \frac{1}{E\gamma} \left[\frac{R_o A}{4kT}\right]^{\frac{1}{2}},$$

where $E\gamma$ is the photon energy. In most cases, the quantum efficiency of the photodiode is limited only by reflection losses and has a value in the range 0.4 - 1.0. The D^* then depends on the zero-bias resistance area product (R_oA) of the junction. This is a sensitive measure of the quality of the material because the junction resistance is inversely proportional to the saturation current, which may be substantially increased by defects that decrease the lifetime in the film.

With large enough values of the resistance-area product, the Johnson noise current becomes smaller than the shot noise due to fluctuations in the rate of arrival of photons from the balckbody background. Under these conditions the D^* becomes background limited with a value

$$D^*(\text{background}) = \frac{1}{E\gamma} \left[\frac{\eta}{2Q_B}\right]^{\frac{1}{2}},$$

where Q_B is the background photon flux in the energy range for which the photodiode is sensitive. The background-limited D^* depends upon the field of view (FOV) of the detector and also quite strongly upon the cut-off wavelength since this influences the value of Q_B. Table 2 shows some representative background-limited detectivities together with the resistance-area products that must be exceeded for background limited operatuion with 180º FOV.

Thus, the demonstration of background-limited D^* at 180º FOV and at some convenient operating temperature (e.g. 77 K) is a convenient criterion for the attainment of material that is good enough to be useful. Further decisions about quality may be based upon the extent to which the D^* increases when the FOV is decreased (i.e. the extent to which R_oA exceeds the value needed for background-limited operation at 180º FOV).

Table 2. Background-Limited Detectivities for Infrared Photodiodes †

Cut-off wavelength (μm)	D^* (Background) (cmHz$^{\frac{1}{2}}$W^{-1})	Equivalent R_oA (ohm cm^2)
3	1.2×10^{12}	3900
4	2.6×10^{11}	115
5	1.2×10^{11}	15
6	7.5×10^{10}	4.1
8	4.6×10^{10}	0.88
10	3.8×10^{10}	0.38
12	3.5×10^{10}	0.22

† These detectivities are calculated for 180º FOV at a wavelength near the cut-off with a typical reflection-loss-limited quantum efficiency of 0.5. The corresponding values of R_oA are those which, at 80 K, give equal contributions from the Johnson noise and the background noise. (This would give detectivities that are reduced by a factor of $\sqrt{2}$ from the background-limited values.)

5. THIN-FILM VI-VI SEMICONDUCTOR PHOTODIODES

While the choice of vacuum deposition technique for thin-film IV-VI semiconductors may be regarded as a matter of taste, the choice of insulating substrates for p-n junction applications is not. With one exception (discussed below) the very extensive body of work with alkali halide substrates has failed to yield p-n junction devices. In contrast, the fluorite structured substrate BaF_2 has given p-n junction devices whose performance is competitive with that of the best bulk crystal devices. Moreover, following the original work with PbTe on BaF_2 at the Ford Laboratory, similar devices have been obtained with a range of IV-VI semiconductors by the Ford group and by several other groups of workers who have used a variety of vacuum deposition techniques. Thus, the key feature is the choice of substrate and the successful development of a p-n junction technology may be attributed to the improvement in the crystal perfection that is described above.

The sole example of significant performance from p-n junction devices made from IV-VI layers grown on alkali halide substrates appears to be the work of Schoolar who worked with PbS on NaCl. For junction formation it was necessary to dissolve away the substrate and use the PbS surface that had been in contact with the NaCl. PbS juncitons that were made by growing p-type PbS on n-type epitaxial PbS layers gave quantum efficiencies around 0.1 and $D*(4~\mu m) = 10^9$ $cmHz^{\frac{1}{2}}W^{-1}$ when operated at 77 K (59). Subsequent work (60) with semitranspartent In metal barriers gave $D*$ (3.8 μm) = 1.5×10^{11} $cmHz^{\frac{1}{2}}W^{-1}$, which is only a factor of two below the background limit (assuming 180° FOV and a typical reflection-loss-limited quantum efficiency of 0.5).

With the observation of high-quality layers of PbTe and (Pb,Sn)Te on BaF_2 substrates (49) it became obviouse to try farication of p-n junction devices with these materials. The first high-performance thin-film IV-VI photodiodes and the first use of the metal barrier technique for IV-VI infrared photodiodes were reported by the Ford group in 1971 (61). Barrier layers of small-work-function metals, such as Pb, had been used previously by Nill *et al.* (62) to make bulk crystal lasers by electrostatically inverting the surfaces of p-type PbTe and (Pb,Sn)Te. Infrared response had been observed from the edges of these devices (63), but optical absorption in the barrier layer had precluded application as photodiodes.*

The energy bands of a thin-film PbTe metal barrier photodiode, following the analysis by Walpole and Nill (65), are shown in Fig. 11.** The BaF_2 substate is conveniently transparent for wavelengths up to about 12 μm. This permits illumination of the junction via the substrate. Taking typical optical absorption lengths of the order of 1 μm and typical minority carrier diffusion lengths of the order of 10 μm, it is evident that a semiconductor layer that is a few μm thick can give both good optical absorption and efficient collection of photogenerated minority carriers.

The original configuration for thin-film PbTe photodiodes (61) was the simple crossed-stripe arrangement that is shown in Fig. 12. When operated at 77 K and 180° FOV, these devices gave background-limited $D*$'s. With reduction of the FOV there was an increase by a factor of four to a Johnson noise limit of $D*(5~\mu m) = 6 \times 10^{11}$

---

* Subsequent work (64) has shown that semitransparent In barriers, similar to those reported by Schoolar, may be used to make high-performance bulk crystal $Pb_{0.8}Sn_{0.2}Te$ photodiodes.

** In most cases we lack an unequivocal demonstration that the thin-film p-n junctions are formed by electrostatic inversion, rather than by shallow diffusion of Pb into, or chalcogen out of, the semiconductor. Despite this uncertainty the junction forming technique is of considerable practical importance.

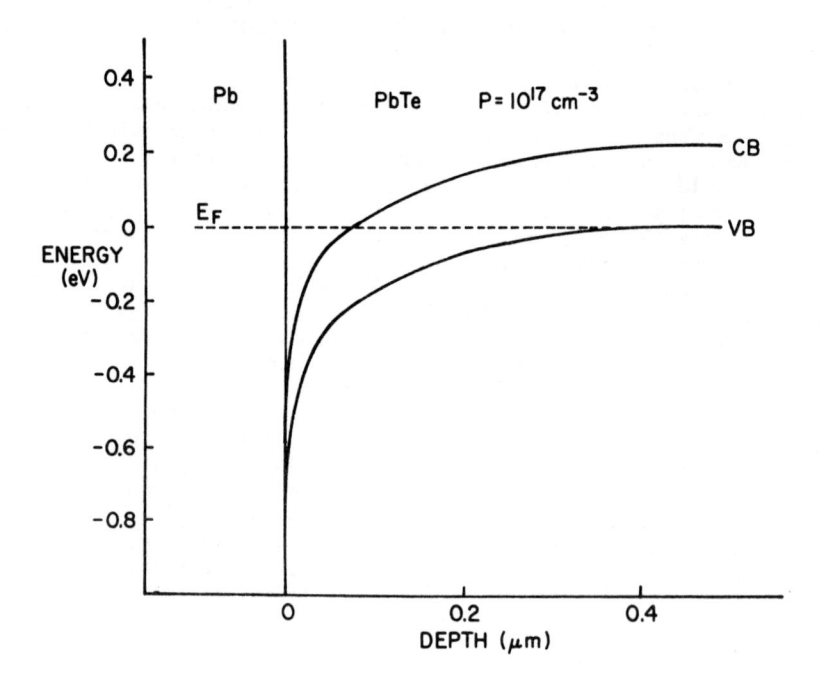

Fig. 11. Calculated band bending due to a Pb layer on
p-type PbTe ($p \approx 10^{17}$ cm^{-3}).

cmHz$^{\frac{1}{2}}$W^{-1}. Subsequent work (54) has given Johnson noise limits as large as 10^{13} cmHz$^{\frac{1}{2}}$W^{-1} for such thin-film PbTe devices at 77 K.

Minor modifications of the crossed-stripe geometry gave the first demonstration of a thin-film injection leaser on an insulating substrate (66) (discussed elsewhere in this review) and also a junction field-effect transistor (67,68). Further studies showed that the performance of the PbTe photodiodes did not depend upon the technique that was used for junciton formation. Essentially similar detectivities were obtained with p-n junctions that were made by proton bombardment (69) and by Sb ion implantation (70) of p-type PbTe layers. * Studies of the stability of the Pb-barrier devices have shown that their performance is retained after repeated cycling to cryogenic temperatures and after more than sufficient baking to outgas vacuum enclosures ($\approx$ 10 hr at 150°C).

Further development from the original thin-film PbTe photodiodes has followed two main lines. First, the use of IV-VI alloys for specific applications in the 3-5 µm atmospheric window and for reduction of the energy gap to permit operation in the 8-12 µm atmospheric window. Secondly, refinement of the rather crude crossed-stripe configuration to permit the delineation of and connection to arrays of small closely-

---

* To date only limited success has been reported with conventional diffusion techniques and with attempts to make p-n junctions by changing the conductivity type of the deposits during film growth. Callender (76) reported $D^*(10$ µm$) = 4 \times 10^9$ cmHz$^{\frac{1}{2}}$W^{-1} at 77 K for Pb$_{0.82}$Sn$_{0.18}$Te devices that were made with a mesa technique applied to 15 µm-thick layers on BaF$_2$. Lopez-Otero et al. (72) reported D^* = 1.4×10^{10} cmHz$^{\frac{1}{2}}$W^{-1} at 77 K for grown-in p-n junctions in epitaxial PbTe on BaF$_2$. In both cases the D^* was an order of magnitude less than the background-limited value that has been exceeded by metal-barrier devices.

spaced detector elements.

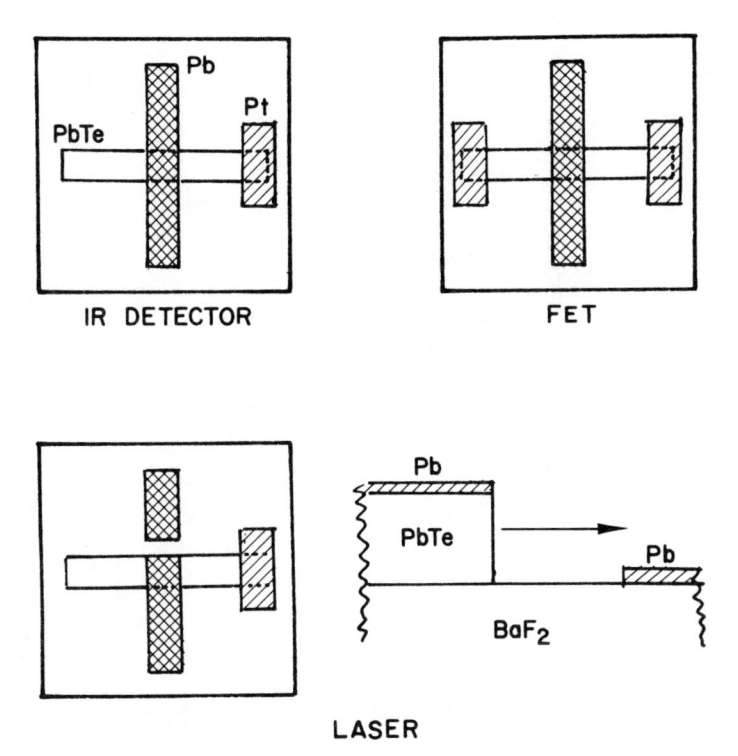

Fig. 12. Crossed-stripe configurations for Pb barrier PbTe devices.

The major advance in thin-film 3-5 μm detectors has been the development of devices whose operating temperature has been increased from 77 K to the intermediate temperature range, 170-200 K, that is suitable for thermoelectric cooling. This permits application in lightweight hand-carried thermal imaging systems. These increased operating temperatures greatly increase the requirements for device quality because of the exponential decrease of the junction resistance with temperature which reduces the Johnson-noise-limited $D*$.

Figure 13 shows the temperature-dependent spectral $D*$ of a thin-film PbTe photodiode (73,74) operated with 180° FOV. The curves show several interesting features. First, there is a sequence of pronounced maxima and minima that arise from interference modulation of the quantum efficiency about the thick-film limit of 0.6. This effect may be used to obtain quantum efficiencies as large as 0.9 for selected wavelengths. Secondly, as the oerpating temperature is increased from 80 K, the $D*$ first increases and then decreases. This effect is due to a change in the dominant noise mechanism. At low temperatures the juncition resistance is sufficiently large that the Johnson noise is much smaller than the background noise. With increased temperature the increase in the energy gap shifts the cut-off to smaller wavelengths and reduces the effective background photon flux thereby increasing the background-limited $D*$. At large enough temperatures the junction resistance decreases to a point where the device becomes Johnson noise limited. Further increase in temperature then reveals the temperature-dependance of this Johnson-noise-limited $D*$.

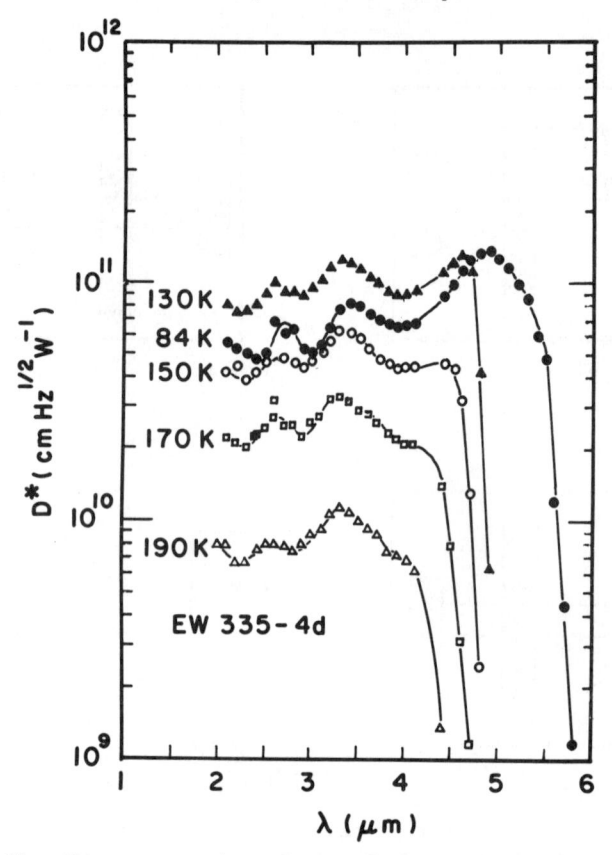

Fig. 13. Temperature dependence of the spectral detectivity of a thin-film PbTe photodiode.

The practically significant feature of Fig. 13 is the attainment of useful D^*'s at temperatures that are suitable for thermoelectric cooling † For many 3-5 μm applications the response of PbTe extends to long enough wavelengths. However, at 170 K the cut-off at about 4.8 μm does not permit full use of the 3.5 μm atmospheric window. This led to consideration of pseudobinary alloys of the Pb chalcogenides that are shown in Table 3.

Table 3. Optical Absorption Edges of Lead Chalcogenides

Material	Optical absorption edge (μm)	
	at 77 K	at 170 K
PbS	3.7	3.3
PbSe	6.9	5.6
PbTe	5.7	4.8

† These results have been confirmed in part by McMahon (75).

Following the demonstration of high performance thin-film Pb barrier PbSe photodiodes (76) the film growth techniques were extended to PbSe$_{0.8}$Te$_{0.2}$ (77) with an optical absorption edge near 5.4 μm at 170 K. This material was used to make the devices whose temperature-dependent spectral D^* are shown in Fig. 14 (73,78). Basically, these results are similar to those obtained with PbTe devices, but with the response extended to cover the whole of the 3-5 μm spectral region. Subsequent work (78) has yielded PbSe$_{0.8}$Te$_{0.2}$ devices with $D^*(5\ \mu m) = 1 \times 10^{11}$ cmHz$^{1/2}$W^{-1} when operated at 170 K with 180° FOV. Under these conditions there are approximately equal contributions from the Johnson noise and the background photon shot noise. This performance appears to exceed that reported for any 3-5 μm photodiode when operated at intermediate temperature. It is perhaps a measure of the maturity of thin-film photodiode technology that the Pb(Se,Te) pseudobinary alloy system was not driven from a bulk crystal application, but was first exploited with the thin-film devices.

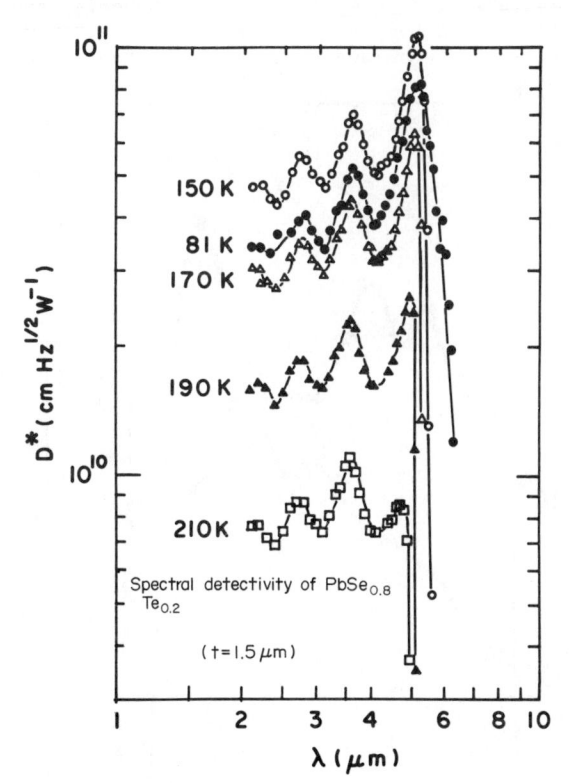

Fig. 14. Temperature dependence of the spectral dtectivity of a thin-film PbSe$_{0.8}$Te$_{0.2}$ photodiode.

A further application of the thin-film technology has been devised by Schoolar et al. (13) who have developed high performance Pb(S,Se) photodiodes. In this work both sides of the BaF$_2$ substrates were coated with Pb(S,Se). One of the layers was used to make Pb barrier photodiodes. The layer on the other side of the substrate had a composition that was adjusted to give a slightly smaller energy gap and served as a long-wavelength pass filter. By careful adjustment of the compositions the response was reduced to a narrow range (∼ 0.2 μm wide at half maximum) that was defined by the cut-on of the filter and the cut-off of the photodiode.

Another major line of development has been the extension of thin-film photodiode response to the 8-12 μm atmospheric window by development of techniques for growth of $Pb_{1-x}Sn_xTe$ (x ≈ 0.2) (48) and $Pb_{1-x}Sn_xSe$ (x ≈ 0.07) on BaF_2 (79). Here, use of the (Pb,Sn)Te alloy system had many precedents in bulk crystal technology, but little success had been reported with bulk crystal (Pb,Sn)Se photodiodes despite early use of this material to establish the existence of the band-crossing phenomenon (29,30).

Figure 15 shows the spectra $D*$'s of two thin-film Pb-barrier (Pb,Sn)Se photodiodes for the 8-12 μm spectral region (81). At 80 K with 180° FOV, such devices are background-limited with resistance-area products up to 2ohm cm^2. With reduction of the FOV, a Johnson noise limit of $D*(10$ μm$) = 8 \times 10^{10}$ $cmHz^{\frac{1}{2}}W^{-1}$ has been obtained. These results compare well with peak $D*$'s of 1×10^{11} $cmHz^{\frac{1}{2}}W^{-1}$ that have been obtained with bulk crystal (Pb,Sn)Te photodiodes at a reduced FOC (82). Approximately background-limited $D*$'s near 10 μm have also been obtained with (Pb,Sn)Te In barrier devices (71,83) at 80 K with 180° FOV. Schoolar and Jenson (14) have also used (Pb,Sn)Se layers on BaF_2 substrates to make narrow response photodiodes by using the same two-layer technique that was described for Pb(S,Se).

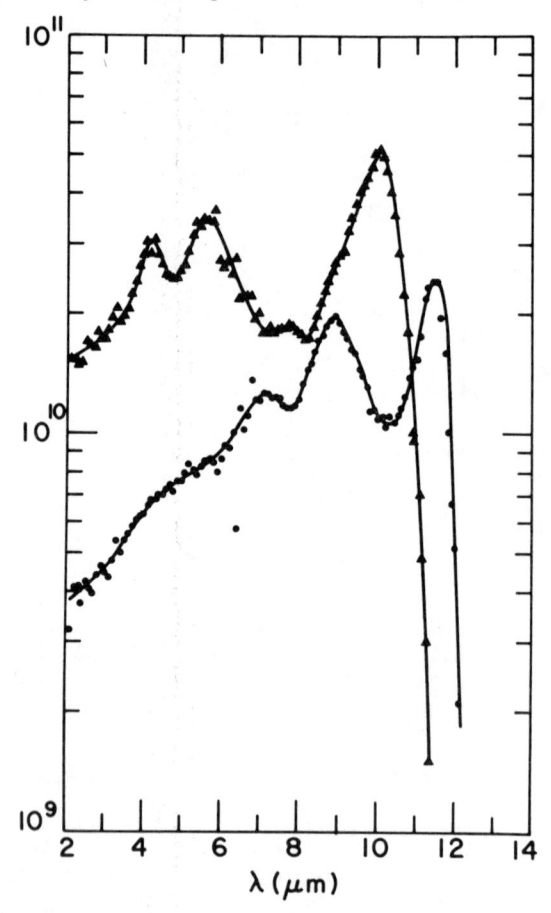

Fig. 15. Spectral detectivities of (Pb,Sn)Se photodiodes at 80 K.

Development of a processing technology for thin-film IV-VI photodiode arrays has been accomplished by Asch and co-workers (84) who have delineated the junction areas with windows in an insulating layer of vacuum-deposited BaF_2. Thus, the basic processes for thin-film array fabrication consist of four successive vacuum depositions. These give the semiconductor, the Pb ohmic contact, the BaF$_2$ insulator, and the Pb barrier, respectively. Mounting of the arrays for illumination through the BaF$_2$ substrates was done with a flip-chip technique that used a composite Cu foil/glass/Cu header for thermal expansion compatibility. Some details of a thin-film array are shown in Fig. 16.

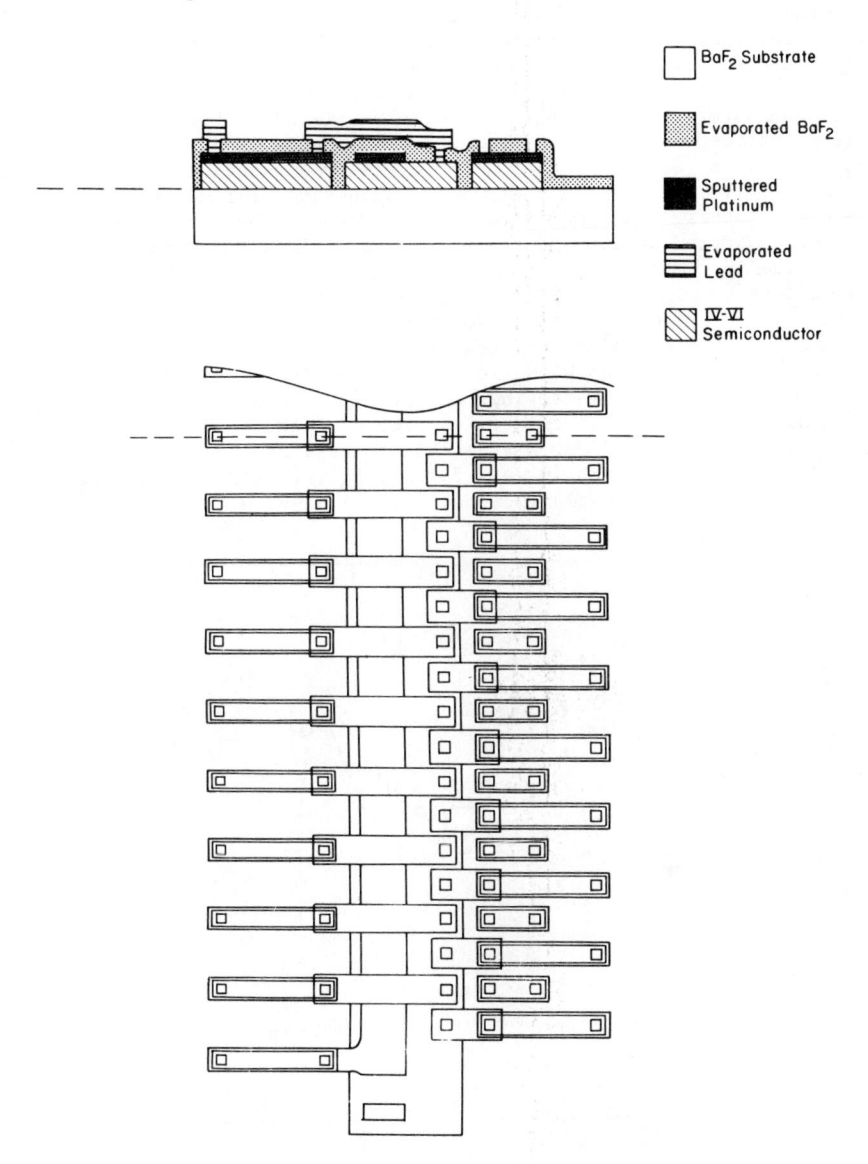

Fig. 16. Structure of a IV-VI semiconductor thin-film photodiode array.

6. UNCONVENTIONAL THIN-FILM IV-VI PHOTODIODES

The preceding section has emphasized the properties of thin-film IV-VI photodiodes whose performance may be compared that of bulk crystal devices to provide an assessment of the perfection of the epitaxial layers. For completeness we now review briefly some more recent developments in which the unique properties of thin films are exploited.

One very characteristic property of the thin-film photodiodes is the interference modulation of the quantum efficiency whose influence is evident in Figs. 13-15. A detailed analysis (85) shows that with layers up to about 5 µm thick the quantum efficiency of IV-VI thin-film photodiodes should be limited only by reflection and transmission losses. The good agreement between the observed quantum efficiency of a PbTe metal barrier device and the calculated reflection-loss limit (RLL) is shown in Fig. 17. The result confirms the attainmnet of bulk-like minority carrier diffusion lengths, and hence lifetimes, in the thin-film material. Exploitation of the interference effects may take two forms. First, by appropriate choice of the layer thickness the RLL at specific wavelengths may be increased to about 0.9 compared with about 0.5 for an uncoated conventional PbTe photodiode. This eliminates the need for antireflection coatings to optimize the detectivity. Secondly by making use of combinations of IV-VI thin films with dielectric layers and metal reflectors the response of the photodiode may be made to resemble that of a combination of a conventional photodiode with an interference filter (85). Such devices may find application for low-cost evaluation of spectral signatures.

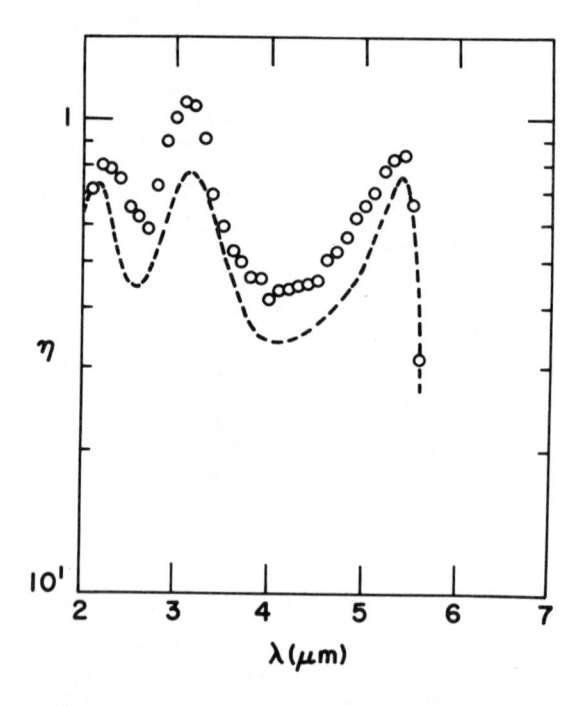

Fig. 17. Comparison of the measured and calculated quantum efficiencies of a thin-film PbTe photodiode at 80 K.

One significant disadvantage of conventional IV-VI semiconductor photodiodes is the is the large junction capacitance (of the order of 1 µFcm^{-2}) that arises from the large dielectric constants of these materials. One approach to capacitance reduction is reduction of the dopant concentration and this has been demonstrated with bulk crystals of (Pb,Sn)Te by Andrews et al. (17). However, since the capacitance of abrupt junction only decreases with the square root of the dopant concentration, this approach requires the growth of material with 10^{15}cm^{-3} carriers or less. The thin-film devices offer an alternative approach that is based on the proximity of the depletion region to the insulating substrate (86). The concept is shown in Fig. 18. With a concentional p-n junction a change in the bias causes a change in the width of the depletion region and thereby a change in the charge that is stored in the depletion region. This gives rise to a dynamic capacitance dQ/dV. However, with the thin-film device the p-n junction may be arranged to have the depletion region extend through to the insulating substrate. Under these conditions the depletion region can only accept or give up charge around its periphery. This leads to a reduction in the capacitance by the factor (peripher X layer thickness)/ (junction area). The capacitance reduction may be achieved by using layers that are thin enough ($\approx$ 0.6 µm) that become pinched off when the depletion region is widened by back bias. For the latter mode it becomes necessary to reduce the 1/f noise that is usually associated with IV-VI photodiodes that are operated in back bias (82). This has been achieved by careful cleaning of the PbTe surface.

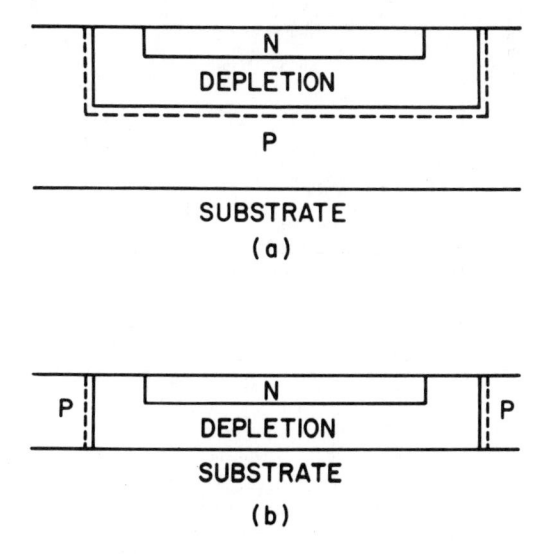

Fig. 18. Conceptual arrangement of the pinched-off photodiode. With increasing back bias, the depletion region edge moves to the position shown as a broken line in (a) for a conventional photodiode and in (b) for a pinched-off photodiode.

The typical bias-dependent performance of a thin-film PbTe pinched off photodiode at 80 K and 180° FOV is shown in Fig. 19. At zero bias the diode with area 6 x 10^{-4} cm^2 has a capacitence of 700 pF. With back bias greater than about 0.15 v the capacitance decreases to a constant value of 70 pF that arises from the contact pads rather than the junction. The 500 K blackbody current responsivity (R_I), which is proportional to the quantum efficiency, is independent of the bias asn the noise remains close to that calculated for fluctuations in the background for back bias up to 0.35 v, after which 1/f noise becomes significant. Thus, the D^* remains at the background-limited value for biases that permit an order of magnitude reduction in the detector capacitance. Larger area photodiodes have shown up to two orders of magnitude reduction in capacitance. It is of interest that no bulk crystal devices exist with properties that are comparable to the thin-film prinched-off photodoides.

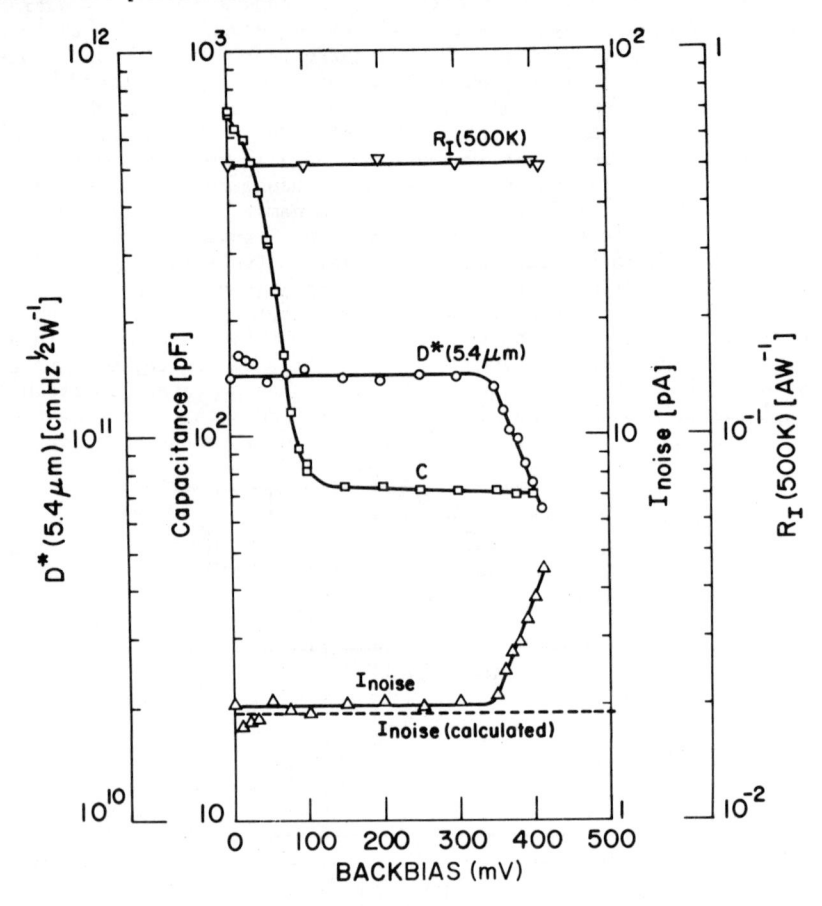

Fig. 19. Bias-dependent performance of an 0.6 μm-thick PbTe photodiode at 80 K and 180° FOV. The diode area is 6×10^{-4} cm^2. The noise was measured at 1 kHz with 10 Hz bandwidth and the broken line shows the background noise current that was calculated from the d.c. background current.

An alternative approach to capacitance reduction using thin-film devices (87,89) is shown in Fig. 20. Here the n-region of a conventional photodiode is replaced by a matrix of small circular n-regions that act as collectors for photogenerated minority carriers from the intervening p-region. The insulating substrate acts as a potential barrier that confines the photogenerated carriers to the vicinity of the collectors, thereby reducing losses by recombination. Efficient collection is obtained with a collector spacing of up to two diffusion lengths (20-30 μm for PbTe). With collector deameters in the range 2-5 μm the reduction in junction area permits more than an order of magnitude reduction in junction capacitance. Figure 21 shows laser scans of such a lateral-collection photodiode (LCP) at different operating temperatures. This shows clearly the increase in collection efficiency as the device is cooled and the diffusion length increases. Under some conditions

the saturation current of IV-VI diodes is generated within the depletion region
and in this case the junction resistance of the LCP is greater than that of a
conventional device with a corresponding increase in the Johnson-noise-limited D^*.
This effect has been used to obtain a 35 K increase in the operating temperature
of PbTe photodiodes. The only comparable bulk crystal work appears to be by
Noreika *et al.* (90) who used collectors with a stripe geometry on bulk crystals
of (Pb,Sn)Te to obtain capacitance reduction by a factor of three.

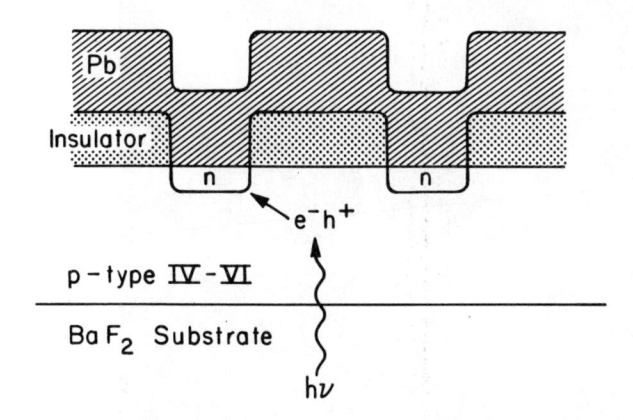

Fig. 20. Schematic arrangement of a lateral-collection photodiode.

7. OPTICAL WAVEGUIDES

An optical waveguide structure consisting of a 6 μm MBE layer of $Pb_{0.92}Sn_{0.08}Te$
grown on a {100} oriented PbTe substrate followed by a 0.5 μm PbTe (cladding)
layer as shown in Fig. 22 has been demonstrated (91). Striped guides were formed
by growing over a SiO_2 mask which was later lifted off to remove the polycrystalline
overgrowth except in the stripes where epitaxial growth occurred. The optical loss
coefficient measured in these guides at 10.6 μm wavelength for light polarized
parallel to the plane of the light-guiding PbSnTe layer (Te polarization) was 7.8
cm^{-1} at room temperature. Since room temperature free carrier absorption has been
well characterized in PbTe (92) and is similar in $Pb_{0.92}Sn_{0.08}Te$ (93), the loss
coefficient of 7.8 cm^{-1} implies an upper limit of about 4×10^{16} cm^{-3} for the free
carrier density of the layer. At 77 K the loss coefficient was too small to measure
but was determined to be ≤ 1.5 cm^{-1}, the smallest optical loss coefficient measured to date in either bulk or epitaxial PbSnTe material. For the orthogonal
polarization (TM), losses were somewhat higher but the increased loss could be
attributed to absorption in the metallization used on the surface of the 0.5 μm
PbTe cladding layer as part of the mounting procedure for the waveguides. The
transmission scans of a CO_2 laser along the path indicated in Fig. 22 are shown in
Fig. 23. The waveguide transmission is strongly reduced by coupling losses into
the narrow guide, compared to transmission through the substrate.

The waveguide results suggest the feasibility of fabricating by MBE monolithic
integrated optical circuits incorporating active elements such as lasers, laser
amplifiers, and detectors. The control of waveguide thickness, which can be
tapered for coupling purposes, using shadowing techniques during growth, † is one
of the advantages MBE offers for this purpose, in addition to the low as-grown

---

† Tapered layers similar to those reported in GaAs MBE layers (94) can easily be
grown in a similar manner in PbSnTe (95).

carrier densities necessary for low loss.

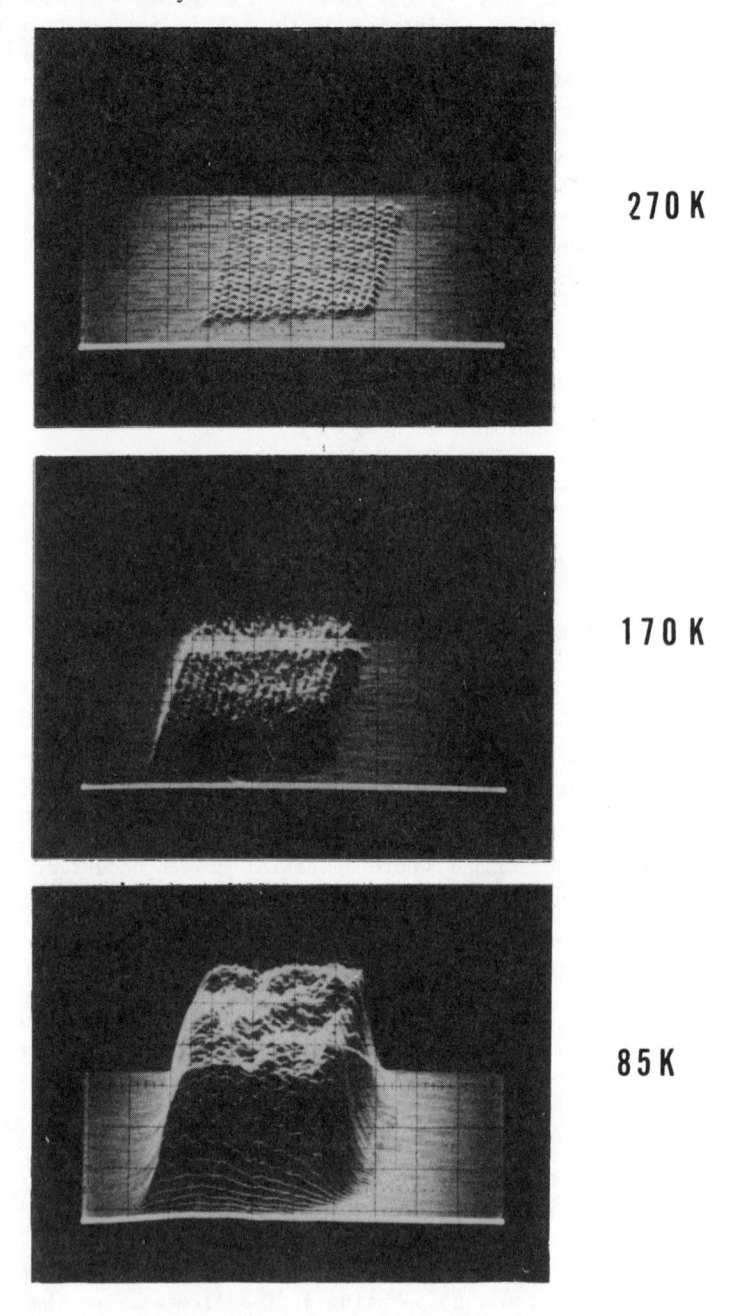

Fig. 21. Laser scans of a 320 μm-square PbTe lateral-collection photodiode at 270 K, 170 K, and 85 K. The 5 μm-diameter collectors are on 20 μm centers. The vertical displacement is proportional to the current response and the instrumental resolution is about 10 μm.

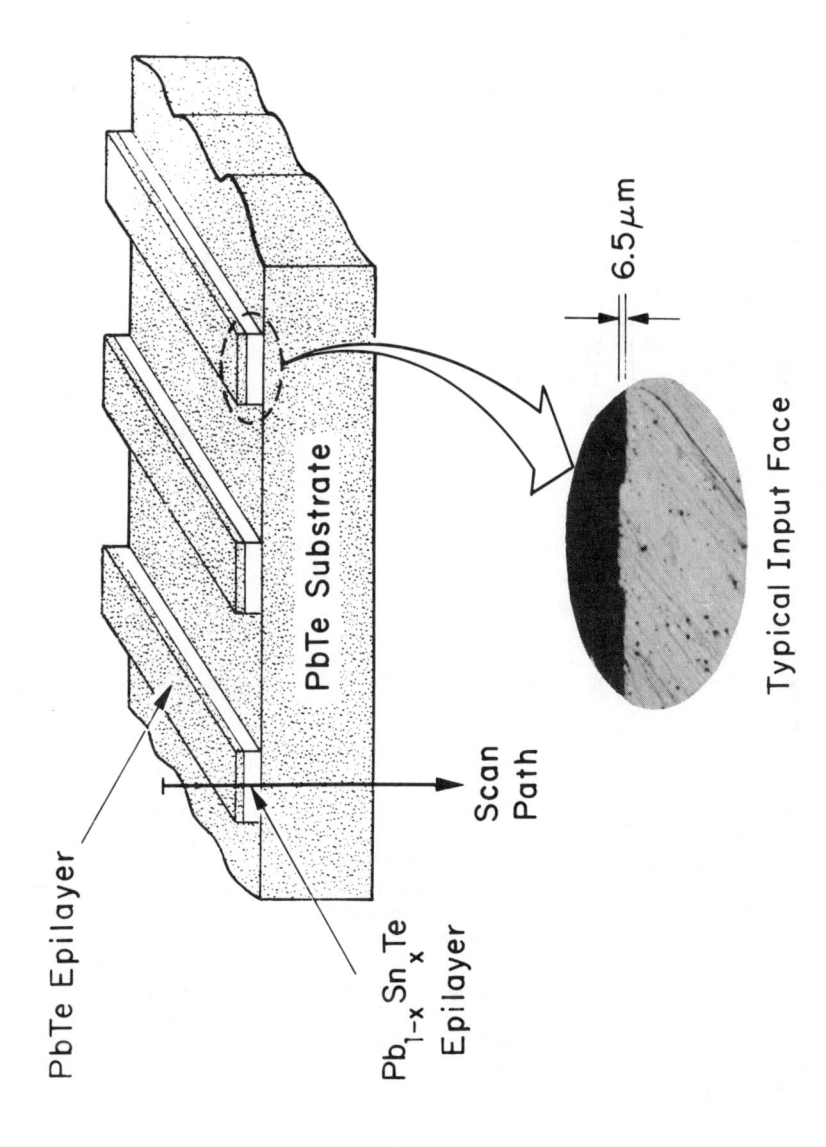

Fig. 22. Schematic of double heterostructure optical waveguides showing the path of scan by a 10.6-μm laser.

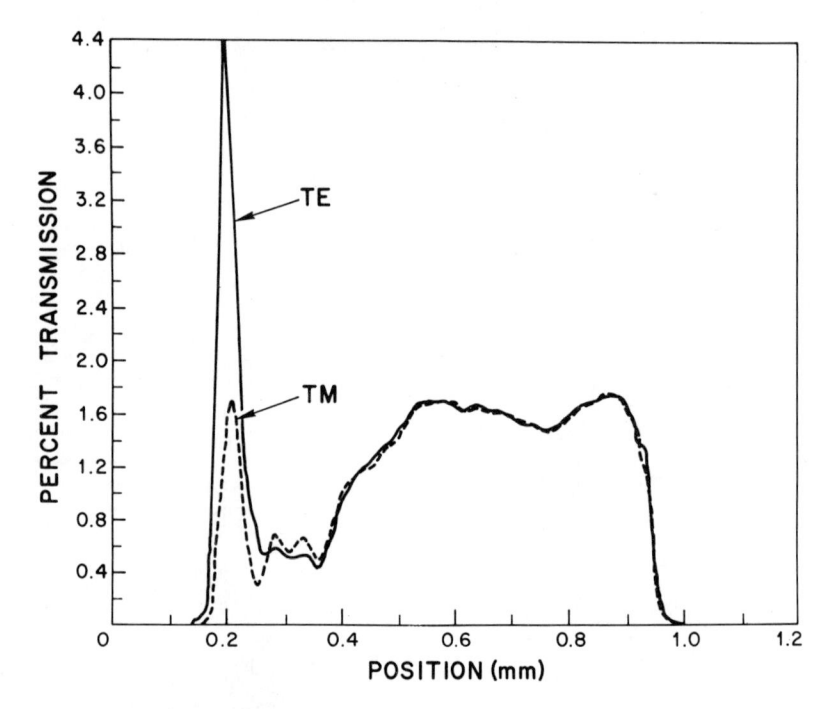

Fig. 23. Transmission scans of TE and TM polarized 10.6 μm laser beams along the path indicated in Fig. 22 for a 4-mm-long waveguide at 77 K. Loss due to coupling is estimated at 95.2 and 92.3% for the TE and TM waveguide modes, respectively. The guiding layer is located at the 0.2-mm position.

8. INJECTION LASERS

Lead-tin chalcogenide injection lasers (laser diodes) have become important research tools for tunable laser spectroscopy (96) in the infrared at wavelengths between about 4 μm and 30 μm (see Fig. 6). Their use as pollution monitoring and gas detection devices (97) has been demonstrated including their potential for remote heterodyne detection (98-100). The motivation for epitaxial growth of these devices has been to achieve cw operation at liquid nitrogen temperature and above, rather than at the low temperature ($\lesssim$ 30 K) previously achieved with diffused p-n junctions in bulk-grown material.* An additional bonus of higher temperature operation is that, as the energy gap changes with temperature, a single device can be used at different temperatures to cover a large spectral region (400-500 cm^{-1}) (101). Fine tuning of a single mode can be achieved at a given temperature by several means but most easily by varying diode current.

The internal change of temperature with current changes the refractive index which shifts the requencies of the longitudinal cavity modes. The continuous tuning range of a mode is typically 1-2 cm^{-1}

---

* In addition to the MBE and hot-wall lasers to be discussed, progress has also been made using liquid-phase epitaxy as mentioned, including the first cw operation of a lead-salt laser above 77K (19).

The use of heterostructures in III-V diode lasers to reduce laser threshold and increase the operating temperature suggested that a similar approach would be successful in the IV-VI materials (102).

8.1. Herterostructure lasers

Figure 24 shows a typical $Pb_{1-x}Sn_xTe$ double heterostructure (DH) laser configuration in which an epitaxial layer (the active region) lies between the substrate and another layer both of which have smaller Sn compositions, larger band-gaps, and smaller refractive indexes than the active region. If a p-n junction is also located within or at an edge of the active region, minority carriers injected at forward bias are confined within this region by the potential barriers resulting from the larger energy gaps at the edges. This carrier confinement is one of the features of DH lasers which, in principle, reduces the threshold current density required for laser action provided that high recombination does not occur at the interfaces. The refractive index variation indicated in the figure forms a dielectric slab waveguide to guide (confine) the optical radiation within the active region where the optical gain is located.

In a single hertostructure (SH) laser, optical and carrier confinement are provided by a change in alloy composition on only one side of the active region. On the other side optical and carrier confinement are provided by a p-n junction or a graded doping profile as in homostructure lasers. As will be seen, in the IV-VI materials strong confinement can be obtained in this way.

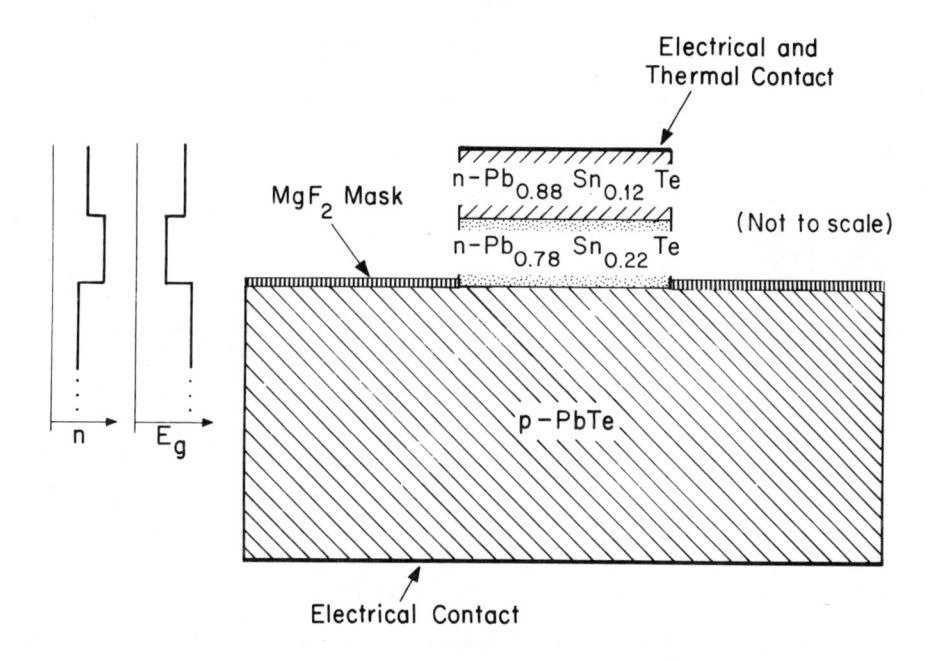

Fig. 24. Schematic diagram of a PbSnTe laser grown with a MgF_2 growth mask. The variation of refractive index n and energy gap E_g are shown at the left.

The first IV-VI heterostructure laser was an unusual variation from the conventional structures discussed above. It consisted of a p-PbTe film grown on a BaF_2 substrate

(66) with a Schottky barrier junction formed by an evaporated Pb strip. Ohmic contact to the film was obtained with sputtered Pt as indicated in Fig. 25. The lateral current flow required by the geometry of this structure leads to lateral voltage drops increasing with the sheet resistance of the film. Hence it is difficult to obtain uniform injection under the strip of Pb. In the devices reported, injection occurred along the edge of the Pb strip only and the active volume of the laser was concentrated within a diffusion length on each side of the edge. The BaF_2 substrate provided strong opitcal and carrier confinement on one side of the film while the Pb Schottky barrier or the interface with air adjacent to it provided optical confinement on the other. There was no significant lateral confinement within the film however. Carrier confinement at the Pb barrier would be characterized by the injection efficiency of the barrier, i.e. the ratio of current flow arising from injected electrons to the toal current flow. The injection efficieny of this type of junction appears to be good at low temperatures (62). Nevertheless, a significant Richardson's current flow could occur in this structure (63), and may prevent high temperature operation.

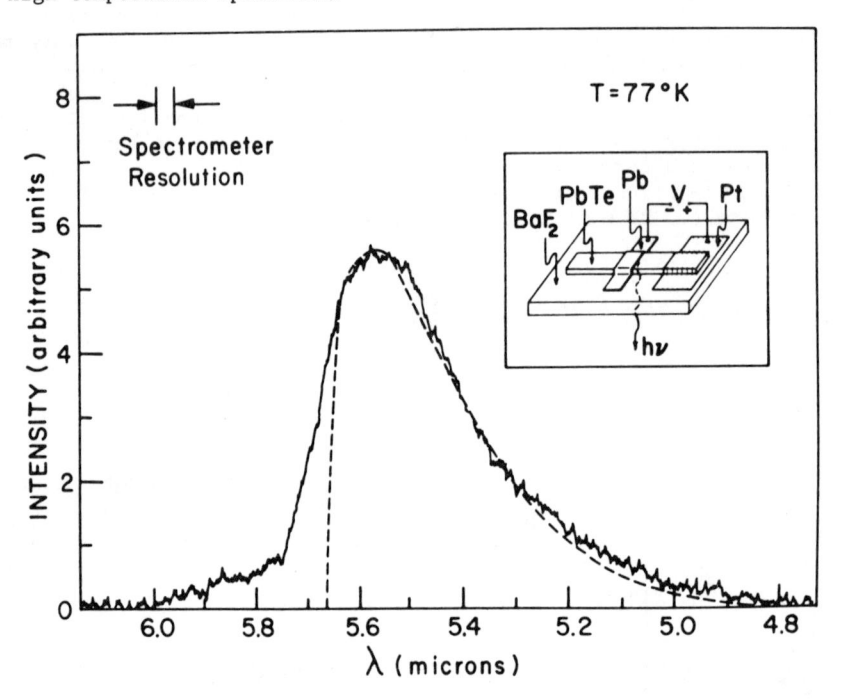

Fig. 25 Spontaneous emission spectrum of a thin-film PbTe laser fabricated as shown in the insert. The theroretical fit (dashed curve) is for k-conserving parabolicband transitions.

Because of the transparent substrate, it was possible to collect a large fraction of the isotropically emitted spontaneous light from these devices so that the spontaneous emission spectrum could be measured. Few well-resolved data of this sort exist otherwise for IV-VI materials. Spectra obtained at 77 K showed a good fit to the theoretical spectrum for direct band-to-band transitions between parabolic bands as shown in Fig. 25.

Although pulsed laser operation was observed in these devices, cw operation was precluded by the lack of adequate heat-sinking. Further improvements in these devices were reported later (103). Post-growth annealing of the films to increase carrier density and thus decrease sheet resistance and an etching technique used to form smooth parallel mirrors resulted in lower thresholds and cw operation was achieved near 10 K. Threshold current densities weree estimated to be about 1 KA/cm^2. The ability to etch smooth parallel mirrors in films grown on BaF$_2$ substrates may be a significant advantage in device fabrication since mirror quality can be a serious problem in PbSnTe lasers because of the poor cleavage properties of the material (104).

In spite of their promise, further development of IV-VI lasers grown on BaF$_2$ substrates has not been pursed to date. Subsequent efforts have made use of the IV-VI materials as substrates where the problem of lateral current flow is avoided and where heat-sinking of the active layer is more straight forward.

The first device of this type was an SH laser in which an n-type layer 3 µm in thickness was grown on a p-type Pb$_{0.88}$Sn$_{0.12}$Te substrate (105). The laser emission wavelength corresponded to the smaller band-gap PbSnTe substrate rather than the PbTe film as expected. The depth of the p-n junction in this device was not established but most likely some diffusion during growth of an n-skin into the substrate on the order of 1 µm deep occurred. A pyrolytically deposited SiO$_2$ growth mask patterned with 50 µm-wide striped openings was used to obtain a stripe geometry. Polycrystalline PbTe grew over the oxide with crystalline growth in the stripes. The oxide did not adhere well to the substrate after growth so that it was not possible to mount the device with the PbTe layer bonded to the heat sink without shorting out the stripe. Instead, the Pb$_{1-x}$Sn$_x$Te bulk side of the device was bonded to the heat sink. Despite the poor heatsinking, cw operation was obtained up to 65 K.

In Fig. 26, the pulsed threshold current density as a function of temperature for a SH laser is compared with data from two low-threshold stripe-geometry homostructure lasers made by diffusion into bulk material. Singificant threshold reduction for the SH laser can be seen. At 77 K the threshold of 780 A/cm^2 is lower by a factor of 3 than any PbSnTe homostructure devices at that temperature reported to date. This result strongly suggests that high recombination at the film-substrate interface does not occur. The improvement of the SH laser over the diffused homostructures is due to the carrier confinement provided by the PbTe layer which prevents diffusion of injected holes to the contact where they would recombine.

A previously unobserved phenomenon was seen in the I-V characteristics of these SH lasers. It has subsequently been observed in a number of injection lasers (106-108) including MBE DH lasers (101). At threshold the I-V characteristic abruptly changes from an exponential to a linear dependence. The change in slope or differential resistance (dV/dI) of an SH laser is shown versus d.c. bias current I at several temperatures in Fig. 27, but the effect is also obvious in the I-V characteristic itself as shown in Fig. 28 for an MBE DH laser at 77 K. The onset of strong stimulated emission at threshold reduces minority carrier liftime so that no further voltage (population inversion) can be applied to the junction and current flow is subsequently limited only by series resistance. As can be seen in Fig. 27 at 30 K and 47 K, there is sometimes a small drop in junction voltage at threshold which may be due to saturable absorption (i.e. less gain is required to maintain lasing action than to reach threshold due to the saturation of absorption processes). The junction voltage saturation implies virtually 100% incremental internal quantum efficiency above threshold (105, 109). It is a consequence of strong carrier confinement withing the optical field (in the lateral direction, by the use of stripe geometry, as well as in the direction transverse to the junction). It should be emphasized that the incremental internal efficiency above threshold is not related to the radiative quantum efficency below threshold.

Sleger *et al.* (110) have reported PbSSe SH lasers grown by the hot-wall technique in which the small band-gap active region is the epitaxial film rather than the substrate.

These devices were only operated pulsed at 12 K, but had threshold current densities of about 200 A/cm^2 comparable to homostructure devices, again implying no serious nonradiative recombination at the hetero-interface nor at misfit dislocations generated by the 0.2% lattice mismatch. The relatively low threshold also indicates that the laser quality of the film in other regards was comparable to bulk-grown crystals.

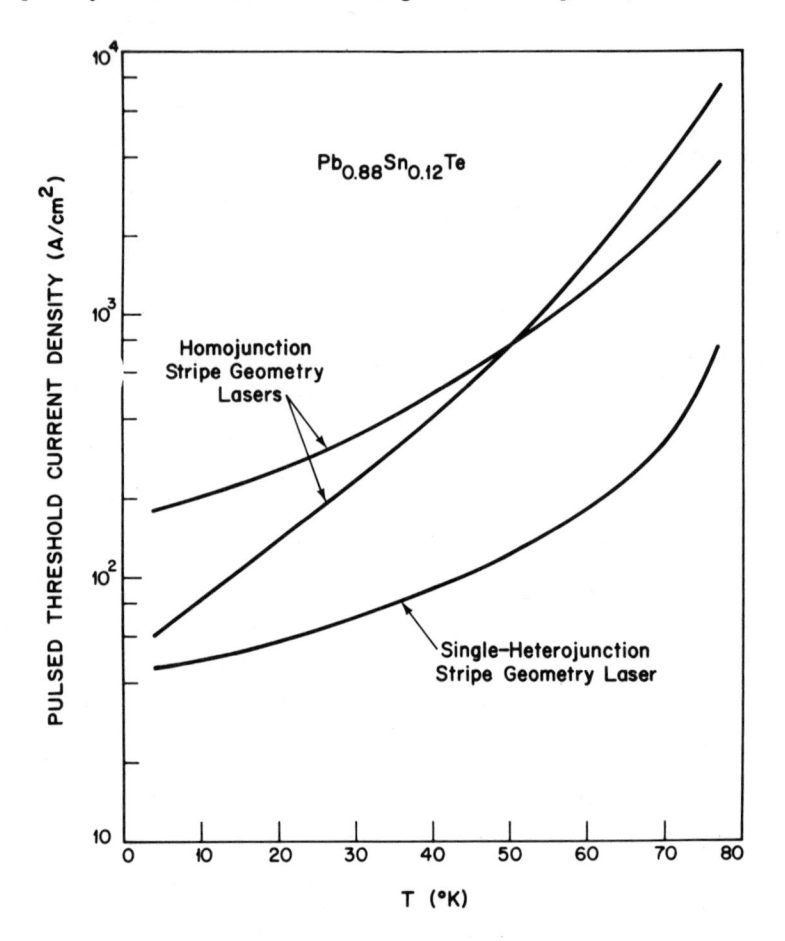

Fig. 26 Pulsed threshold current density of a SH PbSnTe laser as a function of temperature (T) compared with two typical low-threshold diffused-junction bulkcrystal devices.

DH lasers grown by the hot-wall technique were first reported by McLane and Sleger (111) in the PbS$_{1-x}$Se$_x$ alloy stem. Threshold current densities as low as 60 A/cm^2 were obtained for cw operation at 12 K, but no higher temperature data were reported. Preier et al(112) also using the hot-wall technique in the PbS$_{1-x}$Se$_x$ alloy stem, have subsequently obtained DH lasers with low threshold at and above 77 K. These devices were made with striped contacts to form a stripe geometry as seen in the inset of Gig. 29. The pulsed threshold current density as a function of temperature i plotted in Fig. 29 for two DH lasers and, for comparison, a diffused homostructure laser. Above 30 K and 60 K, respectively, the two DH lasers show lower threshold current densities than the diffused laser. These devices were capable of cw operation up to 96 K. Higher temperature operation could have been achieved with a thinner PbS top layer which at 5 μm thickness represents a large thermal impedance (the thermal conductivity of PbS at 100 K is about 0.07 W cm^{-1} K^{-1}) (112). Later, similar devices with 1.5 μm thick top layers were operated cw up to 120 K (113).

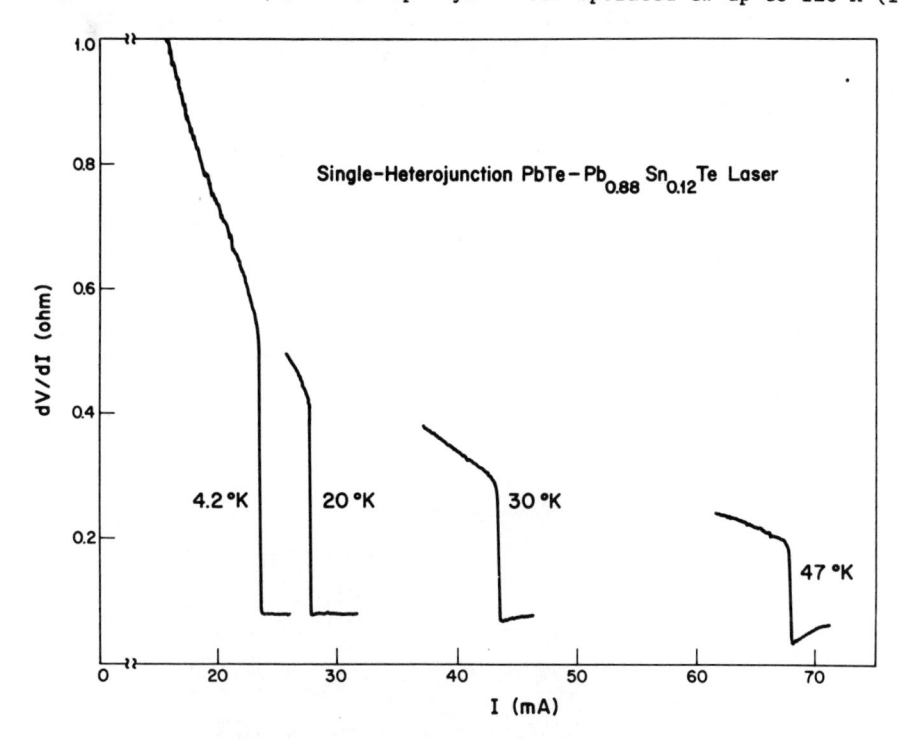

Fig. 27 The incremental resistance (dV/dI) of a Sh PbSn Te laser vs current I at several temperatures. The abrupt decrease in resistance occurs at cw leaser threshold.

At currents well above threshold, laser modes were seen over a sifficiently broad spectral region that slight inhomogeneity or grading of the PbSSe alloy composition seemed indicated.

In the hot-wall PbSSe results discussed above, control of carrier type and density was achieved by an independently controlled Se vapor source. While similar control of type can be achieved in conventional MBE (79), it is difficult to achieve high p-type concentrations needed for low resistance contacts. In addition, diffusion of the p-n junction can be rapid even at the low temperatures of MBE growth as mentioned earlier. McLane and Sleger (111) estimated their control of junction position to be no better than 1 μm. For these reasons, the PbSnTe DH lasers grown by MBE have been doped with foreign impurities as discussed in Section 2. Relatively high concentrations can be easily achieved and junction diffusion is expected to be slower for foreign dopants, as mentioned there.

The PbSnTe DH lasers were grown in a stripe geometry using a striped growth mask of MgF$_2$ with stripe widths of 10 to 50 μm. Substrates were Tl-doped PbTe ($\sim 10^{19}$ cm^{-3}). Active regions 1.5 μm thick were grown using Pb$_{0.78}$Sn$_{0.22}$Te, Bi-doped to achieve about 6 X 10^{17} cm^{-3} n-type concentration. Cap layers 0.75 μm thick were grown using Pb$_{0.88}$Sn$_{0.12}$Te, Bi-doped to achieve 2 X 10^{18} cm^{-3} n-type concentration. These devices were capable of cw operation up to 114 K (101). However, threshold current densities were large particularly for low temperatures as seen in Fig. 30, where the threshold and the wavelength of emission are plotted vs temperature.

Subsequent devices have had somewhat lower thresholds, ranging between 2 and 6 KA cm^{-2}, but remaining large compared to diffused homostructure devices, the SH PbnTe devices, or the PbSSe devices discussed above. It can be easily determined from the wavelength of the emission at threshold and from the I-V characteristics such as that shown in Fig.28, that the voltage across the junction at threshold is within a few mV of the voltage required to establish population inversion. Since the current flow required to reach population inversion is clearly the dominant contribution to the threshold current (115), the high thesholds cannot be explained by optical losses. Instead there appears to be a large component of current due to surface leakage or nonradiative recombination. While misfit dislocations or interface recombination cannot be ruled out as the source of this current, the evidence of the SH lasers and the PbSSe lasers does not support hsi view (41), nor do the results obtained with epitaxial homostructure devices discussed below. Other possibilities include surface leakage, nonradiative recombination associated with the high levels of foreign impurities present (116), and nonradiative recombination due to a non-equilibrium, high concentration of native defects which have been proposed to explain low mobility in unannealed MBE films (49) of PbSnTe on BaF$_2$.

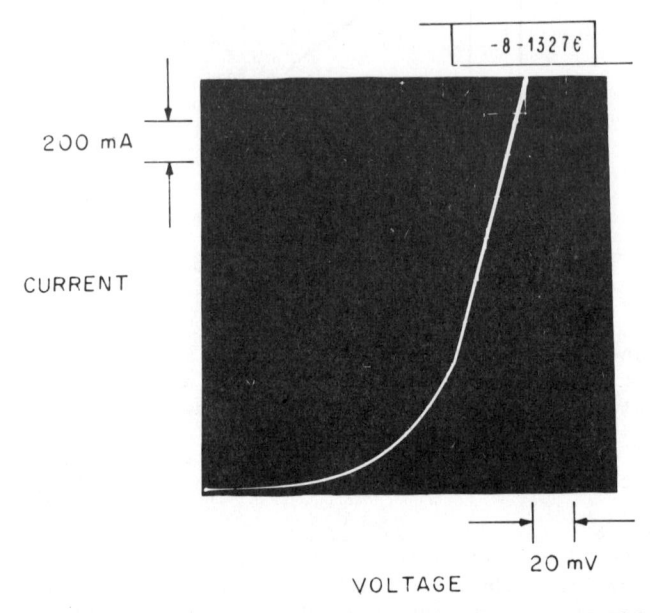

Fig. 28 The current-voltage characteristic of a DH PbSnTe laser at 77 K. Note the linear relationship above threshold at I = 600 mA.

The high threshold of the PbSnTe DH lasers grown by MBE do not significantly limit their performance since cw operation can be achieved for current densities over 20 KA cm^{-2}. Adequate power and mode quality for high-resolution spectroscopy at 80 K has been demonstrated (117). The control of the thickness of the cap layer and the active layer, made possible by MBE, is a crucial factor in obtaining good performance since, as mentioned, the poor thermal conductivity of the lead-tin chalcogenides causes poor heat-sinking for junctions more than a few μm deep.

The MBE techniques used for the PbSnTe DH lasers have also proven extremely reproducible in numerous growth runs and the yield of devices with cw operation above 77 K from each wafer is high.

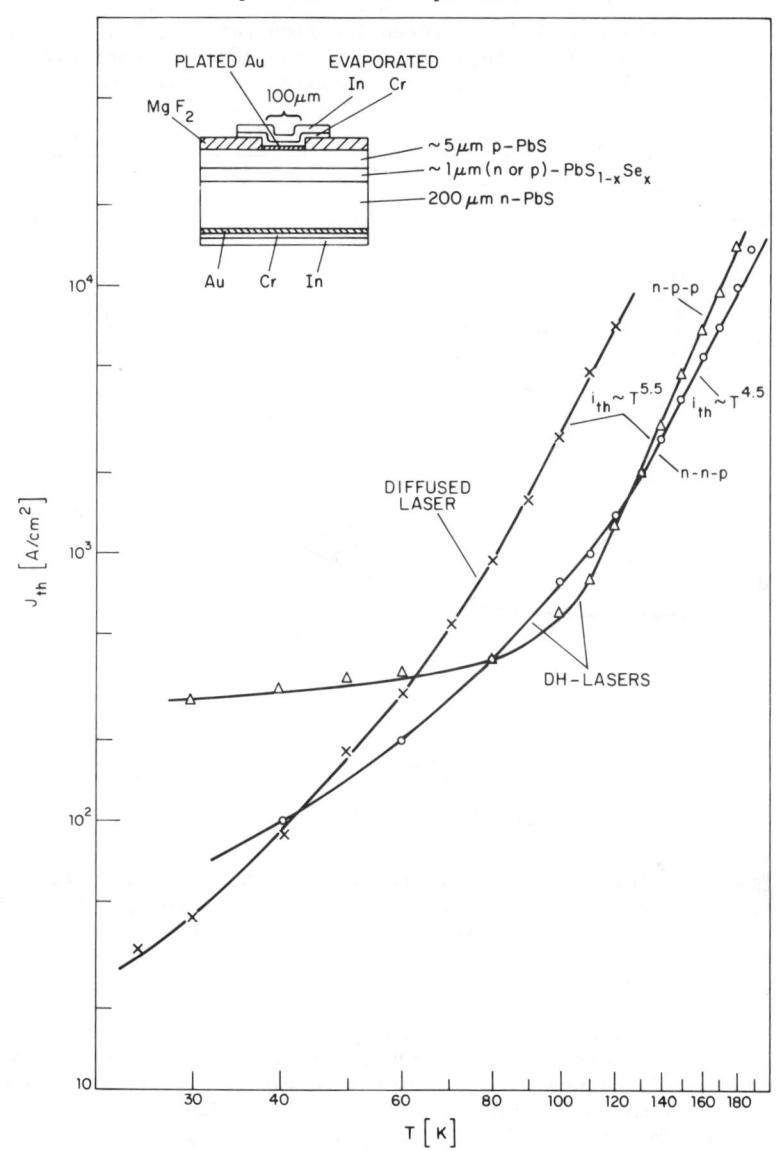

Fig. 29 Comparison of pulsed threshold current densities of
DH PbSnTe lasers with a bulk crystal diffused-junction
laser as a function of temperature (from Ref. 112).

No other lasers grown by MBE or hot-wall techniques have been reported. There are several papers of interest in the Russian literature (116,118,119). However, no high performance at higher temperatures was mentioned and the growth techniques were not always specified but appear to involve vapor epitaxy.

A disappointing aspect of the MBE and hot-wall heterostructure lasers reported to date is that their external quantum efficienies have been low (less than a percent or two). Typical maixmum output power has been on the order of a few hundred μW in PbSSe and about 100 μW in PbSnTe.* Much higher external efficiencies and output powers have been achieved in bulk diffused-junction lasers (104, 106). Although the powers obtained from heterostructures are more than adequate for most applications, they are marginal for local oscillator applications in heterodyne detection.

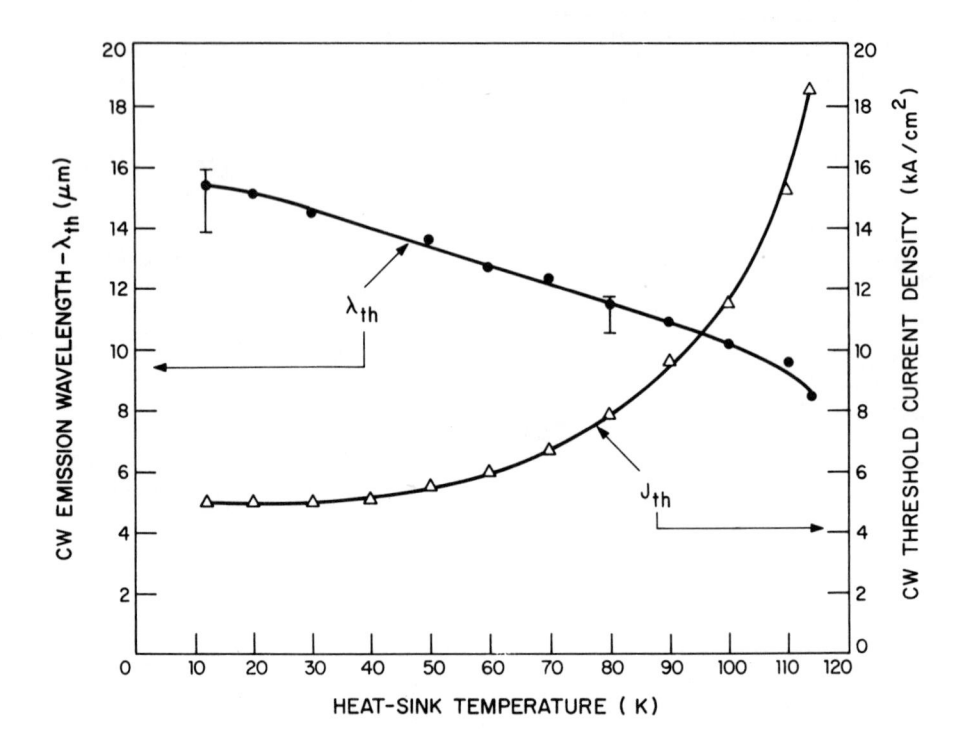

Fig. 30. Threshold current density and emission wavelengths of a DH PbSnTe laser as a function of heat-sink temperature. The longest wavelength observed pulsed compared to the cw threshold wavelength gives a measure of the temperature increase in the active region.

The low external efficiency cannot generally be attributed to low internal efficiency since the DH PbSnTe lasers which show junction-voltage saturation at threshold also have low external efficiency. Poor output coupling and scattering losses at the end mirrors due to poor mirror quality (104) and due to mode-coupling effects (121) may play a large role in the low efficiency. Also scattering losses and free-carrier absorption in the active region may be much larger in laser devices than in the waveguide structures discussed in section 4. The laser waveguides are much thinner, which leads to increased scattering due to roughness of the waveguide walls (103), and carrier densities in the lasers are much higher**. Optimized doping and the use of separate optical and carrier confinement structures might significantly improve external efficiency.

---

* An early DH $Pb_{0.88}Sn_{0.12}Te$ laser (120) with an undoped active region 6 μm thick and an undoped PbTe cap 1 μm thick grown on a Tl-doped PbTe substrate gave cw output power up to 9 mW. These devices were not capable of cw operation above $\simeq$ 30 K, however.

** Additional absorption above the classical free-carrier absorption may be present near the band edge in heavily doped materials (93,122).

8.2. Homostructure lasers

Strong optical and carrier confinement can be achieved in the lead-tin chalcogenides with properly controlled carrier-concentration profiles (123). Because of the small effective masses of both electrons and holes, relatively large built-in potentials (band bending) occur at interfaces between low and high carrier concentration. Also large free-carrier contributions to the dielectric constant occur because of the small effective masses and the small energy gaps or long wavelengths of operation.

Diffused-junction lasers have been operated cw recently up to 61 K using concentration gradients for confinement (124). The diffusion is very critical and difficult to control since the junction depth must be kept very small for heat-sinking. Epitaxial techniques are better suited for controlling concentration and depth of the junction. Epitaxial homostructure devices capable of cw operation to 100 K have been reported (125) grown by both MBE and LPE having n^+-n-p^+ doping profiles. Bi-doped active regions ($n \simeq 5 \times 10^{17}$ cm^{-3}) 1.5 µm thick followed by Bi-doped capping layers ($n^+ \simeq 5 \times 10^{18}$ cm^{-3}) 0.75 µm thick were grown on Tl-doped substrates ($p^+ \simeq 10^{19}$ cm^{-3}).

The MBE homostructures were grown in PbTe and $Pb_{0.78}Sn_{0.22}Te$, both compositions operating cw to 100 K. Threshold current densities were comparable to the DH lasers but output powers were somewhat better. Figure 31 shows the cw spectra of a PbTe homostructure at 15 and 102 K with 2.4 mW and 0.64mW output, respectively, predominantly in a single mode. The maximum output observed from $Pb_{0.78}Sn_{0.22}Te$ homostructures was ∿ 0.4 mW at 15 K and ∿ 0.12mW at 80 K.

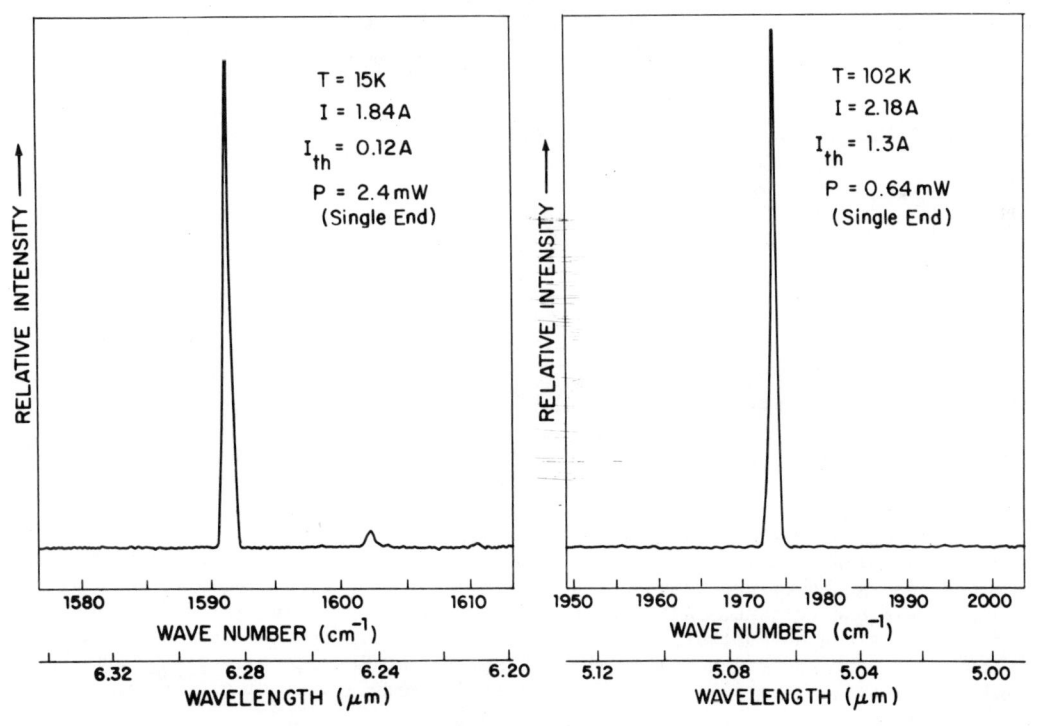

Fig. 31. Spectra of a MBE homojunction PbTe laser at 15 and 102 K showing the relatively high single-mode output powers.

Since thresholds are comparable to DH lasers and external efficiencies are still rather low, the homostructure results again imply that the lattice mismatch present in otherwise similar heterostructure lasers plays little role in limiting their performance. Nevertheless, if lattice matching should prove important, MBE homostructures seem to offer the advantage of higher temperature operation without the lattice mismatch problem.

8.3. Distributed feedback lasers

Figure 32 shows schematically the structure of a distributed feedback (DFB) laser which has been fabricated in MBE-grown PBSnTe layers (126, 127). The device is essentially the same as the DH lasers discussed above except that feedback is provided by Bragg reflection from the periodic grooves (grating) etched into the top layer rather than by reflections from the two ends of the device. The 800-Å insulating MgF_2 film was patterned as shown so that only one end of the mesa stripe was electrically contacted leaving an uncontacted length with high optical absorption to prevent reflective feedback from the other end of the mesa. The grating and the deeper grooves isolating the mesa stripe were obtained using photolithographic techniques and rf sputter etching.

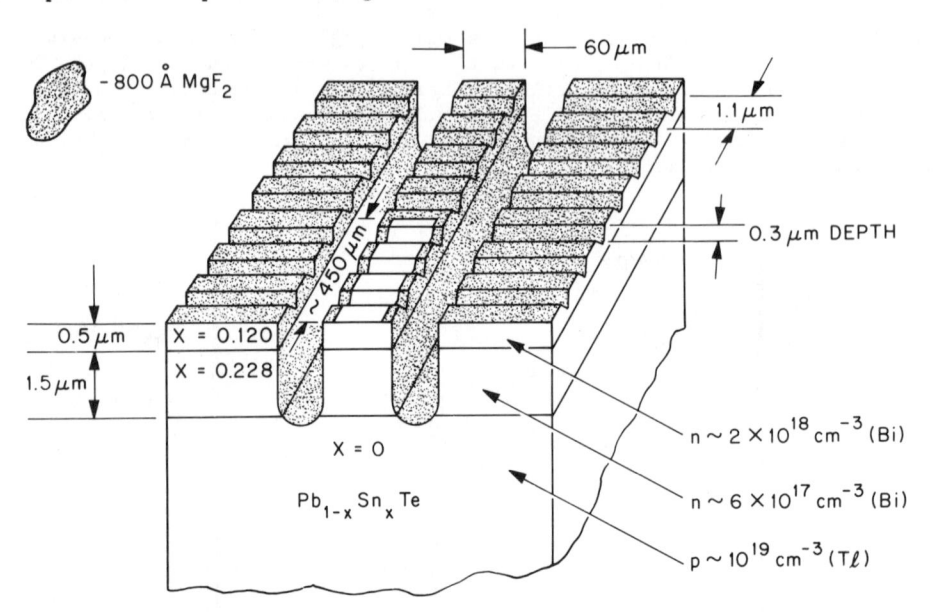

Fig. 32. Schematic of DFB double-heterostructure PbSnTe laser without metallization.

One advantage of the DFB laser is that is has significant mode discrimination, i.e. the longitudinal spatial modes of the cavity have different thresholds. The PbSn Te DFB lasers showed predominantly single-mode cw operation over a wide range of temperature and current. Another advantage of DFB lasers is that the frequency of the laser can be controlled by the grating periodicity since oscillation can occur only near the frequency at which Bragg reflection occurs. These advantages make DFB lasers potentially the ideal laser for many applications.

The successful fabrication of the DFB structure shown in Fig. 32 is due largely to the ability to accurately control the thickness of the top layer to <0·1 µm. It is a device for which MBE techniques are ideally suited. The scanning electron micrographs shown in Fig. 33 depect two views of grating of 1·1 µm periodicity with grooves 0.3 µm deep. The grooves must be close to the active layer to obtain good coupling for Bragg reflection. They must not, however, penetrate the active layer or carrier confinement would be lost. The cap layer was grown 0.5 μm thick, so that control of etching parameters as well as layer thickness if critical for good performance.

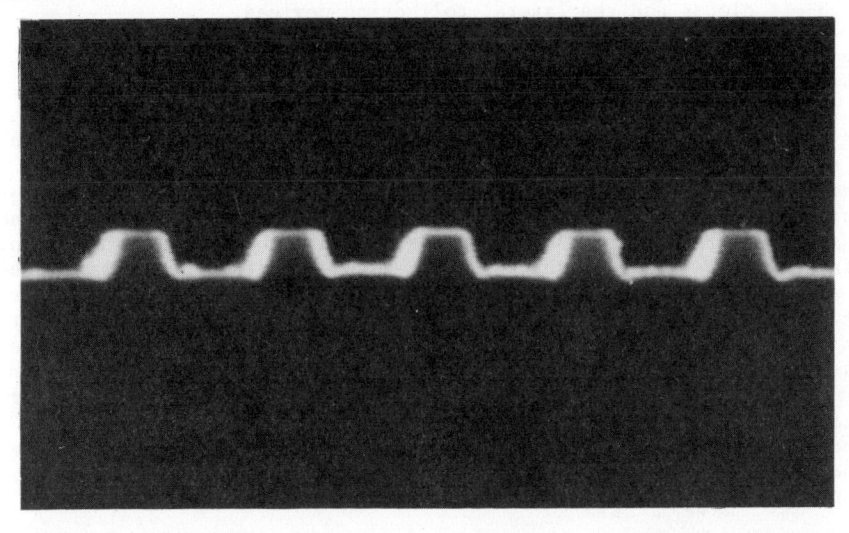

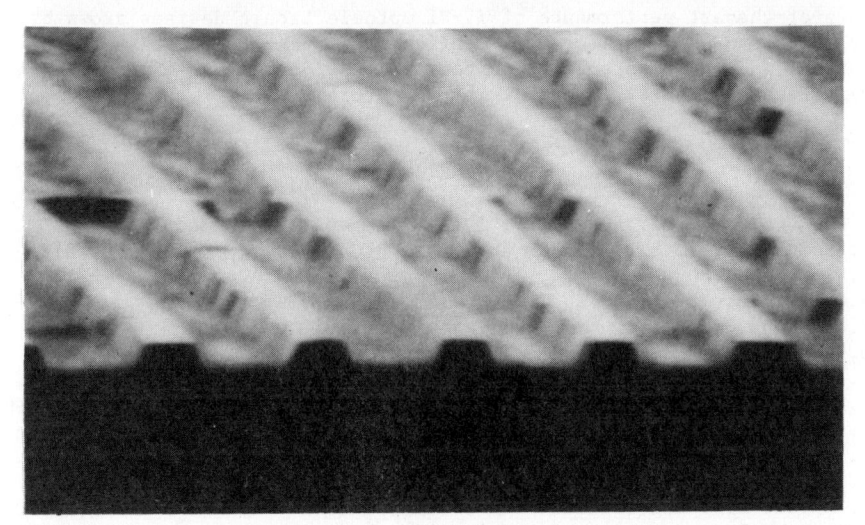

Fig. 33. Scanning electron micrographs of a grating with 1.1-μm periodicity, sputter etched to a depth of 0.3 μm in PbSnTe.

9. CONCLUSIONS

The optoelectronic device results which have been reviewed here demonstrate that high-quality devices, competitive in performance with bulkcrystal devices, can be achieved in IV-VI MBE films. Indeed there are technological advantages and performance features discussed here which have not been demonstrated in bulkcrystal devices. These included the option of backside illumination of detector devices with transparent substrate; the compatibility with planar technology for small closely-spaced elements and for applications in integrated optics; and the control of layer thicknesses which makes possible laser devices with higher temperatures of cw operation, DFB lasers devices utilizing thin-film optics, and (because of the confinement provided by the proximity of an insulating substrate) devices such as the pinch-off diodes and the lateral collection devices.

While MBE techniques have led to advances in device technology, it is also true that the device results so obtain have given important information on the quality of MBE films. The quantum efficiencies and current-voltage characteristics of junction devices give evidence about minority carrier lifetime, generation-recombination mechanisms, and interface recombination. The generally high quality with respect to these properties found in the detector devices is not found in many of the laser structures discussed, as evidenced by the high thresholds. Also optical absorption, which can be low in passive MBE waveguides, appears to be high in the thin, heavily doped layers of the laser structures, as evidenced by the low external quantum efficiencies. Although the device results suggest approaches for improvements and further progress is likely, there is need for more work to correlate crystal perfection (e.g. native defect concentrations, foreign impurities, misfit dislocations, interface states, etc.) with the properties significant for device performance, i.e. carrier recombination and optical losses. Criteria for good device performance are not established even for bulk crystals and better knowledge would greatly simplify the process of optimizing the growth and the design and fabrication of devices. To this end, more sophisticated MBE systems with in-situ analysis equipment such as Auger and HEED instrumentation might prove fuitful.

The state-of-the-art performance of IV-VI optoelectronic devices grown by MBE is already more than adequate for many applications. Because of their demonstrated applications and their potential for new applications, IV-VI MBE devices are likely to remain technologically important in the future.

REFERENCES

1. H. Hinterberger, *Z. Phys.* **119**, 1 (1942).
2. L. Sosnowski, J. Starkiewicz and O. Simpson, *Nature, Lond.* **159**, 818 (1947).
3. A.J. Elleman and H. Wilman, *Proc. Phys. Soc. Lond.* **61**, 164 (1948).
4. S.G. Parker, *J. electrochem. Soc.* **123**, 920 (1976).
5. R.K. Pandy, *Solid St. Commun.* **15**, 449 (1974).
6. L.R. Koller and H.D. Coghill, *J. electrochem. Soc.* **107**, 973 (1960).
7. P. Hudock, *Trans. Am. Inst. Mech. Engrs.* **239**, 338 (1967).
8. R.F. Bis, J.R. Dixon and J.R. Lowrey, *J. vac. Sci. Technol.* **9**, 226 (1972).
9. A. Lopez-Otero and L.D. Haas, *Thin Solid Films* **23**, 1 (1974).
10. K. Duh and H. Preiser, *Thin Solid Films* **27**, 247 (1975).
11. I. Kasai, J. Hornung and J. Baars, *J. electron. Mater.* **4**, 299 (1975).
12. I. Kasai, D.W. Bassett and J. Hornung, *J. appl. Phys.* **47**, 3167 (1976).
13. R.B. Schoolar, J.D. Jensen and G.M. Black, *Appl. Phys. Lett.* **31**, 620 (1977).
14. R.B. Schoolar and J.D. Jensen, *Appl. Phys. Lett.* **31**, 536 (1977).
15a. L.L. Chang and R. Ludeke, in: *Epitaxial Growth, Part A*, (ed.) J.W. Matthews, Academic Press, New York, San Francisco, London (1975).
15b. J.N. Zemel, in: *Solid State Surface Science*, (ed.) M. Green, Dekker, New York (1969).
16. A. Lopez-Otero, *Thin Solid Films* **49**, 3 (1978).
17. A.M. Andrews J.A. Higgins, J.T. Longo, E.R. Gerner and J.G. Pasko, *Appl. Phys. Lett.* **21**, 285 (1972).
18. C.C. Wang and S.R. Hampton, *Solid-St. Electron.* **18**, 121 (1975).
19. S.H. Groves, K.W. Hill and A.J. Strauss, *Appl. Phys. Lett.* **25**, 331 (1974).
20. T.F. Tao, I. Kasai and J.F. Butler, *J. vac. Sci. Technol.* **6**, 918 (1969).
21. E. Krikorian *J. vac Sci. Technol.* **12**, 186 (1975).

22. R.F. Porter, *J. chem. Phys.* **34**, 583 (1961).
23. R.F. Brebrick and A.J. Strauss, *J. chem. Phys.* **40**, 3230 (1964).
24. R.F. Brebrick and A.J. Strauss, *J. chem. Phys.* **41**, 197 (1965).
25. R.F. Bis, *J. vac. Sci. Technol.* **7**, 126 (1970).
26. W.C. Chan, *Infrared Phys.* **14**, 177 (1974).
27. D.A. Northrop, *J. electrochem. Soc.* **118**, 1365 (1971).
28. M.C. Finn and J.N. Walpole, unpublished.
29. A.J. Strauss, *Phys. Rev.* **157**, 608 (1967).
30. R.W. Bicknell, *Infrared Phys.* **17**, 57 (1977).
31. R.W. Bicknell, *Infrared Phys.* **18**, 133 (1978).
32. T.C. Harman, in: *Physics of IV-VI Compounds and Alloys*, (ed.) S. Rabii, Gordon & Breach, London (1974).
33. J.N. Walpole and R.L. Guldi, in: *Physics of IV-VI Compounds and Alloys*, (ed.) S. Rabii, Gordon & Breach, London (1974).
34. G. Dionne and J.C. Woolley, *Phys. Rev.* **B6**, 3898 (1972).
35. D.L. Smith and V.Y. Pickhardt, *J. electron. Mater.* **5**, 247 (1976).
36. H. Holloway, D.K. Hohnke, R.L. Crawley and E. Wilkes, *J. vac. Sci. Technol.* **0**, 00(000).
37. J.W. Matthews, *Phil. Mag.* **6**, 1347 (1961).
38. J.W. Matthews, *Phil. Mag.* **8**, 711 (1963).
39. K. Yagi, K. Takayanagi, K. Kobayaski, and G. Honjo, *J. Crys. Growth* **9**, 84 (1971).
40. J.N. Walpole, S.H. Groves, A.R. Calawa and T.C. Harman in: Solid State Reseach Report, Lincoln Laboratory, M.I.T. (1974:4), p.13 DDC AD-A002773/0.
41. R.W. Bicknell, *J. vac. Sci. Technol.* **14**, 1012 (1977).
42. J.N. Zemel, J.D. Jensen and R.B. Schoolar, *Phys. Rev.* **140A**, 330 (1965).
43. J.H. Myers, R.H. Morriss and R.J. Deck, *J. appl. Phys.* **42**, 5578 (1971).
44. H. Holloway, in: *Use of Thin Films for Physical Investigations*, (ed.) J.C. Anderson, Academic Press, New York (1966).
45. R.S. Allgaier and B.B. Houston, Jr., *Proc. Int. Conf. Phys. Semiconductors*, p. 172, Exeter, 1962, Institute of Physics and Physical Society, London, (1962).
46. E.M. Logothetis and H. Holloway, in: *Physics of IV-VI Compounds and Alloys*, (ed.) S. Rabii, Gordon & Breach, London (1974).
47. J.P. Donnelly, T.C. Harman and A.G. Foyt, *Apply. Phys. Lett.* **18**, 259 (1971).
48. H. Holloway, E.M. Logothetis and E. Wilkes, *J. appl. Phys.* **41**, 3453 (1970).
49. H. Holloway and E.M. Logothetis, *J. appl. Phys.* **42**, 4522 (1971).
50. T.O. Fahrinre and J.N. Zemel, *J. vac. Sci. Technol.* **7**, 121 (1970).
51. H.M. Manaserit and W.I. Simpson, *J. electrochem. Soc.* **122**, 444 (1975).
52. V.L. Lambert, *J. appl. Phys.* **46**, 2304 (1975).
53. D.K. Hohnke, H. Holloway and M.D. Hurley, *Thin Solid Films* **38**, 49 (1976).
54. H. Holloway, unpublished data.
55. A. Lopez-Otero, *J. appl. Phys.* **48**, 446 (1977).
56. R.S. Allgaier and W.W. Scanlon, *Phys. Rev.* **111**, 1029 (1958).
57. G.A. Antcliffe and J.S. Wrobel, *Mater. Res. Bull.* **5**, 747 (1970).
58. C.M. Wolfe and G.E. Stillman, *Appl. Phys. Lett.* **18**, 205 (1971).
59. R.B. Schoolar, *Appl. Phys. Lett.* **16**, 446 (1970).
60. R.B. Schoolar, *J. vac. Sci. Technol.* **9**, 225 (1972).
61. E.M. Logothetis, H. Holloway, A.J. Varga and E. Wilkes, *Appl. Phys. Lett.* **19**, 318 (1971).
62. K.W. Nill, A.R. Calawa, T.C. Harman and J.N. Walpole, *Appl. Phys. Lett.* **16**, 375 (1970).
63. K.W. Nill, J.N. Walpole, A.R. Calawa and T.C. Harman, in: *The Physics of Semimentals and Narrow-Gap Semiconductors*, (eds) D.L. Carter and R.T. Bate, Pergamon Press, Oxford (1971).
64. H.B. Morris, R.A. Chapman, R.L. Guldi and S.G. Parker, paper presented at *IRIS Detector Specialty Group Meeting*, Fort Monmouth, 1975.
65. J.N. Walpole and K.W. Nill, *J. appl. Phys.* **42**, 5609 (1971).
66. H. Holloway, W.H. Weber, E.M. Logothetis, A.J. Varga and K.F. Yeung, *Appl. Phys Lett.* **21**, 5 (1972).

67. H. Holloway, W.J. Johnson and A.J. Varga, unpublished data.
68. H. Holloway, in: *Physics of IV-VI Compounds and Alloys*, (ed.) S. Rabii, Gordon & Breach, London (1974).
69. E.M. Logothetis, H. Holloway, A.J. Varga and W.J. Johnson, *Appl. Phys. Lett.* 21, 411 (1972).
70. J.P. Donnelly and H. Holloway, *Appl. Phys. Lett.* 23, 682 (1973).
71. R.E. Callender, paper presented at *IRIS Detector Specialty Group Meeting*, Washington, D.C., 1973.
72. A. Lopez-Otero, L.D. Haas, W. Jantsch and K. Lischka, *Appl. Phys. Lett.* 28, 546 (1976).
73. Contract DAAKK 2-73-C-0225, Final Report, Febuary 1974.
74. H. Holloway, K.F. Yeung, A.E. Asch and D.A. Gorski unpublished data.
75. T.J. McMahon, Paper presented at *IRIS Detector Specialty Group Meeting*, El Toro, 1974.
76. D.K. Hohnke and H. Holloway, *Appl. Phys. Lett.* 24, 633 (1974).
77. D.K. Hohnke and M.D. Hurley, *J. appl. Phys.* 47, 4975 (1976).
78. H. Holloway, D.K. Hohnke, K.F. Yeung, A.E. Asch and D.A. Gorski, unpublished data.
79. D.K. Hohnke and S.W. Kaiser, *J. appl. Phys.* 45, 892 (1974).
80. J.F. Butler, A.R. Calawa and T.C. Harman, *Appl. Phys. Lett.* 9, 427 (1966).
81. D.K. Hohnke, H. Holloway, K.F. Yeung and M.D. Hurley, *Appl. Phys. Lett.* 29, 98 (1976).
82. W.H. Rolls and D.V. Edolls, *Infrared Phys.* 13, 143 (1973).
83. D.P. Mathur, *Opt. Engng.* 14, 351 (1975).
84. A.E. Asch, D.A. Gorski, P.I. Zapella and H. Holloway, paper presented at *IRIS Detector Specialty Group Meeting*, Fort Monmouth, 1975.
85. H. Holloway, *J. appl. Phys.* (to be published).
86. H. Holloway and K.F. Yeung, *Appl. Phys. Lett.* 30, 210 (1977).
87. Contract DAAK02-75-C-0106, Final Report, June 1976.
88. H. Holloway, M.D. Hurley and E.B. Schermer, *Appl. Phys. Lett.* 32, 65 (1978).
89. H. Holloway, *J. appl. Phys.* (to be published).
90. A.J. Noreika, M.H. Francombe, W.J. Taei, R.N. Ghoshtagore and J.L. Wentz, paper presented at *IRIS Detector Specialty Group Meeting*, Colorado Springs, 1977.
91. R.W. Ralston, J.N. Walpole, T.C. Harman, and I. Melngailis, *Appl. Phys. Lett.* 26, 64 (1975).
92. A.J. Strauss, *J. Nonmetals* 1, 133 (1973).
93. C. Pickering, *Solid St. Commun.* 22, 395 (1977).
94. J.L. Merz, R.A. Logan, W. Wiegmann, and A.C. Gossard, *Appl. Phys. Lett.* 26, 337 (1975).
95. J.N. Walpole, unpublished.
96. E.D. Hinkley, K.W. Nill, and F.A. Blum, in: *Topics in Applied Physics* Vol. 2, *Laser Spectroscopy of Atoms and Molecules*, (ed.) H. Walter, Springer-Verlag, Berlin, Heidelberg, New York (1976).
97. E.D. Hinkley, R.T. Ku, and P.L. Kelley, in: *Topics in Applied Physics*, Vol. 14, *Laser Monitoring of the Atmosphere*, (ed.) E.D. Hinkley, Springer-Verlag, Berlin Heidelberg, New York (1976).
98. M. Mumma, T. Kostiuk, S. Cohen, D. Buhl, and P.C. Von Thuna, *Nature, Lond.* 253, 514 (1975).
99. M.A. Frerking and D.J. Muehlner, *Appl. Opt.* 16, 526 (1977).
100. R.T. Ku and D.L. Spears, *Opt. Lett.* 1, 84 (1977).
101. J.N. Walpole, A.R. Calawa. T.C. Harman and S.H. Groves, *Appl. Phys. Lett.* 28, 552 (1976).
102. K.J. Sleger, G.F. McLane, and D.L. Mitchell, in: *Physics of IV-VI Compounds and Alloys*, (ed.) S. Rabii, Gordon & Breach, London (1974).
103. W.H. Weber and K.F. Yeung, *J. appl. Phys.* 44, 4991 (1973).
104. J.N. Walpole, A.R. Calawa, R.W. Ralston, and T.C. Harman, *J. appl. Phys.* 44, 2905 (1973).
105. J.B. Walpole, A.R. Calawa, R.W. Ralston, T.C. Harman, and J.P. McVittie, *Appl. Phys. Lett.* 23, 620 (1973).

106. R.W. Ralston, J.N. Walpole, A.R. Calawa, T.C. Harman, and J.P. McVittie, *J. appl. Phys.* 45, 1323 (1974).
107. T.L. Paoli and P.A. Barnes, *Appl. Phys. Lett.* 28, 714 (1976).
108. K.J. Linden, K.W. Nill, and J.F. Butler, *IEEE J. Quantum Electron.* QE-13, 720 (1977).
109. H.S. Sommers, Jr., *Appl. Phys. Lett.* 32, 547 (1978).
110. K.J. Sleger, G.F. McLane, U. Steom, S.G. Bishop, and D.L. Mitchell, *J. appl. Phys.* 45, 5069 (1974).
111. G.F. McLane and K.J. Sleger, *J. electron, Mater.* 4, 465 (1975).
112. H. Preier, M. Bleicher, W. Riedel, and H. Maier, *Appl. Phys. Lett.* 28, 669 (1976).
113. H. Preier, M. Bleicher, W. Riedel, and H. Maier, *J. appl. Phys.* 47, 5476 (1976).
114. D.K. Honke and S.W. Kaiser, *J. appl. Phys.* 45, 892 (1974).
115. L.R. Tomasetta and C.G. Fonstad, *IEEE J. Quantum Electron.* QE-11, 384 (1975).
116. L.N. Kurbatov, A.D. Britov, S.M. Karavaev, G.A. Kalyuzhnaya, M.I. Nikolaev, O.V. Pelevin, B.G. Girich, and T.F. Terekhovich, *Sov. J. Quantum Electron.* 7, 236 (1977).
117. P.F. Moulton, D.M. Larsen, J.N. Walpole, and A. Mooradian, *Opt. Lett.* 1, 51 (1977).
118. L.N. Kurbatov, A.D. Britov, S.M. Karavaev, Y.I. Gorina, G.A. Kalyuzhnaya, and P.M. Starik, *Sov. J. Quantum Electron.* 5, 1137 (1975).
119. A.D. Britov, S.M. Karavaev, G.A. Kalyuzhnaya, Y.I. Gorina, A.L. Kurbatov, K.F. Kiseleva, and P.M. Starik, *Sov. J. Quantum Electron.* 6, 1217 (1976) and 6, 1384 (1976).
120. J.N. Walpole, R.W. Ralston, A.R. Calawa, and T.C. Harman, paper presented at the *IEEE International Semiconductor Laser Conference*, Atlanta (1974).
121. R.W. Davies and J.N. Walpole, *IEEE J. Quantum Electron.* QE-12, 291 (1976).
122. S.L. McCarthy, W.H. Weber, and M. Mikkor, *J. appl. Phys.* 45, 4907 (1974).
123. P.N. Asbuk, Ph.D. Thesis, M.I.T., 1975 (unpublished).
124. W. Lo, *IEEE J. Quantum Electron.* QE-13, 591 (1977).
125. J.N. Walpole, S.H. Groves, and T.C. Harman, *IEEE Trans. Electron Devices* ED-24, 1214 (1977). Also, Solid State REsearch Report, Lincoln Laboratory, M.I.T. (1977:4), p. 3, DDC AD-A052463.
126. J.N. Walpole, A.R. Calawa, S.R. Chinn, S.H. Groves, and T.C. Harman, *Appl. Phys. Lett.* 29, 307 (1976).
127. J.N. Walpole, A.R. Calawa, S.R. Chinn, S.H. Groves, and T.C. Harman, *Appl. Phys. Lett.* 30, 524 (1977).

THE AUTHORS

Dr. H. Holloway

Professor J. N. Walpole

Dr. Holloway was educated at Birkbeck College, University of London, where he received his B.Sc. (1956) and Ph.D. (1959) degrees in Chemistry. His research interests have included x-ray and electron diffraction, epitaxy — particularly of compound semiconductors — and the physics of thin-film devices. Currently, he is studying the properties of optoelectronic devices made with thin films of IV-VI Semiconductors.

James Walpole was granted a B.S.E.E. degree in 1961 from Duke University, and then attended the Massachusetts Institute of Technology as a National Science Foundation Fellow, receiving his M.S. degree in 1962, and his Ph.D. degree in 1966, both in the Department of Electrical Engineering. His major studies were in the areas of solid state physics and semiconductor devices. He was appointed Assistant Professor of Electrical Engineering at M.I.T. in 1966, and Associate Professor in 1971. Since 1972 he has been a staff member in the Applied Physics Group of the M.I.T. Lincoln Laboratory, where he is currently involved with development of GaInAsP heterostructure lasers. His previous research interests have centered around the physics of the lead-tin chalcogenide semiconductors, including studies of magnetoplasma wave propagation, diffusion, hot-electron conduction, metal-semiconductor contacts, and, in recent years, various aspects of injection lasers in these materials. He lives in Concord, Massachusetts with his wife and two children.

INTEGRATED OPTICAL DEVICES FABRICATED BY MBE

R. D. Burnham and D. R. Scifres

XEROX PARC, Palo Alto, CA 94304, U.S.A.

(Submitted March 1979)

1. INTRODUCTION

Approximately 10 years ago, S. E. Miller (1) at Bell Telephone Laboratories outlined a proposal for taking various discrete and bulky optical components, which for many years had been spread out over a large optical bench, and miniaturizing them on a common rigid substrate. The incorporation of such discrete devices into an optical circuit would eliminate mechanical and thermal stability problems, along with the need for costly and heavy optics and expensive bulky optical tables. The size of these miniaturized components at least in one or two dimensions needed to be only as large as the optical carrier wave (0.4-10 µm) and thus were amenable to mass production using standard photolithographic techniques. It was proposed at that time that the miniaturization of such components onto a common substrate, a chip, be called an integrated optical circuit in view of its electrical counterpart the integrated circuit. Today, several components of such an integrated optical circuit have indeed been integrated onto a common substrate and have been demonstrated to achieve improved stability, small size, light weight, and low power requirements. In fact, some chips consisting of arrays of lasers now appear to be a reality for optical communications applications and no doubt other applications will follow.

Fiber optic communications has provided the primary stimulus for integrated optics research. Integrated optics coupled with fiber optics is expected to find application in communications where the data rates exceed 300 Mb/sec and where spatial multiplexing is extensively used. This application may require single mode, CW laser sources modulated at rates up to 10 GHz, low loss single mode fibers and GHz response photodectectors.

Monolithic and hybrid are two approaches in which one could make integrated optics. The hybrid approach uses several materials which are chosen to optimize each of the components. This obviously can present serious problems since generally processes used to fabricate the different devices might be incompatible.

The monolithic approach uses one material which works well for all the components. Since the best laser, modulator, waveguide and detector are not necessarily found in one material, compromises will have to be made. Fortunately, however, some semiconductors appear to be suitable for monolithic optical integration. In fact, because of the dual role that direct bandgap semiconductors play in both generating and detecting optical radiation, such materials are virtually the only ones suitable for monolithic integration.

2. GaAs AND RELATED ALLOYS

One of the most promising and most worked on semiconductor materials is GaAs and its alloys with columns III-V. GaAs technology has progressed to the point that high quality fairly large area substrates (4-5 cm in diameter) can be produced. Substrates of other III-V compounds of similar size are either not available or the cost is prohibitive. The strong concentration on the GaAs family is due to its versatility in terms of its electrical and optical properties and to the fact that it is the only materials system in which all the important optical functions — light generation, guiding, modulation, and detection — have been achieved. This makes it possible, in principle, to make monolithic integrated optical circuits out of the GaAs family with suitable addition of impurities, just as integrated circuits are made of silicon.

An attractive feature of GaAs for integrated optics is that its alloys with other compounds of column III-V elements have refractive indexes and bandgaps that vary with the amount of the other compound added. For example, the refractive index of $Ga_{1-x}Al_xAs$ at a wavelength of 0.9 μm decreases almost linearly with x from 3.59 for x = 0 to 2.97 for x = 1 (see Fig. 1). It is clear then that waveguides can be made by epitaxial growth of layers of different x on top of each other. For such structures to have good device performance, it is important that the lattice constants of adjoining layers be well matched. If that is not the case, metallurgical imperfections, such as dislocations, will occur at the interfaces, giving rise to such deleterious features as optical scattering centers and electron-hole recombination centers. $Ga_{1-x}Al_xAs$ is particularly attractive because the lattice constant for GaAs is 5.653 Å and for AlAs, 5.661 Å; thus, the largest possible difference between lattice constants in this family is only 0.14%

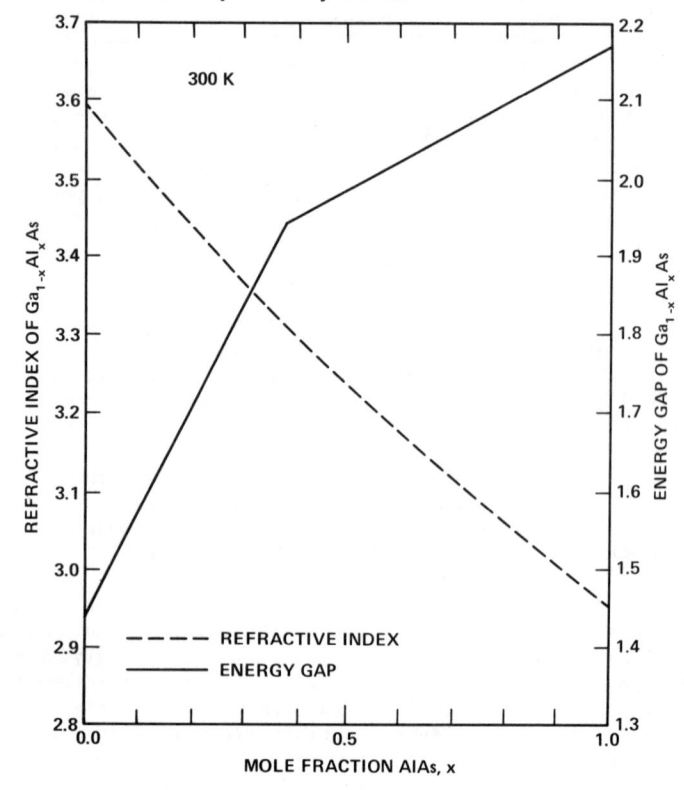

Fig. 1. Variation of refractive index and bandgap of $Ga_{1-x}Al_xAs$ with x.

The interconnection of optical components via waveguides is the central problem of monolithic integrated optics for two main reasons: first, the minute cross sections of the waveguides (10^{-8}–10^{-7} cm^2) require strict dimensional tolerances, and second, optical circuits consisting of active and passive waveguide components need adjacent regions with different material properties such as composition, bandgap, index of refraction, doping, etc. These requirements can only be satisfied by taking the properties and limitations of the crystal growth techniques into account.

3. MOLECULAR BEAM EPITAXY

There are three growth techniques which could be used for integrated optical circuits based on single crystal semiconductors, they are: liquid phase epitaxy (LPE), chemical vapor phase epitaxy (VPE), and molecular beam epitaxy (MBE). Even though more work has been done with optical integration by LPE, MBE appears to be the most promising growth technique because of its overall versatility. VPE has recently received considerable recognition because of advances made by Dupuis and Dapkus (2) using metalorganics but it still lacks the overall flexibility required to make extremely complicated monolithic integrated optical circuits compared to MBE.

The initial pioneering work on MBE was done by A. Y. Cho at about the same time that S. E. Miller proposed integrated optics. Today one of the major research goals is to merge these two technologies so as to produce a variety of integrated optical circuit elements.

The advantages of growing GaAs/Ga$_{1-x}$Al$_x$As integrated optical circuits by MBE are many: (a) Since the growth is performed in ultra high vacuum with the availability of in situ diagnostic instruments, one can be assured that the desired surface conditions have been reached before commencement of the growth; (b) the MBE is controlled by kinetics rather than by a diffusion process and thus allows one to grow layers with practically any predetermined doping profile (3) and variation of composition (4,5); (c) large uniform layers (3 cm x 3 cm) with featureless surfaces may be grown by MBE; (d) the low epitaxial temperature (580°C) may prevent dopant diffusion both within the epitaxial layer and from the substrate into the epitaxial layer; and (e) the slow growth rate (1 μm/hr) allows for essentially monolayer <5 Å thickness control (6-9). The major problem with MBE for integrated optics applications has been centered on the growth of high quality Ga$_{1-x}$Al$_x$As.

This review article will therefore deal mainly with the various MBE techniques that have been used towards the goal of making monolithic integration in the GaAs/Ga$_{1-x}$Al$_x$As system. We refer the reader to other review articles (10,11) and books (12,13) on a broader coverage of integrated optics which discuss other growth techniques and considerably more theory than will be presented here.

4. WAVEGUIDES

First, consider the optical waveguide, which is perhaps the most basic of all integrated optics components. Most of these guides are planar, i.e., those with rectangular geometry. The guiding property can be understood by studying the heterostructure illustrated in Fig. 2. As shown, the waveguide consists of a sandwich of three different material compositions, each of which is transparent to the propagating light-wave. A necessary condition for waveguiding in these heterostructure layers is $n_2 > n_1, n_3$, where n is the refractive index. The waveguiding property can be thought to result from the phenomenon of total internal reflection Consider a ray of light starting in material 2 and propagating toward the 2-3 interface. If the angle between the normal to the interface and the direction of lightwave propagation θ_{23} is larger than the critical angle for total internal reflection

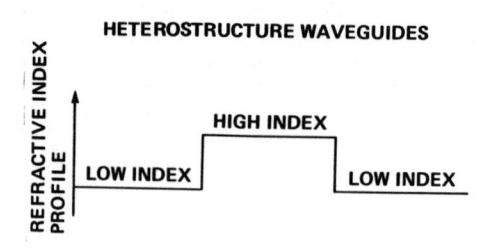

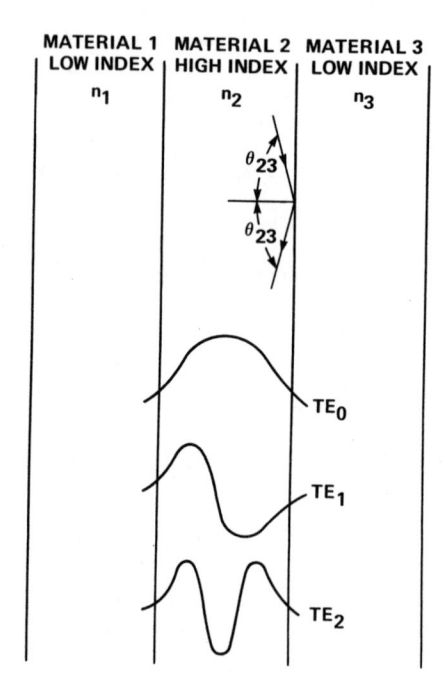

Fig. 2. Schematic diagram of a heterostructure waveguide.

[$\theta_c = \sin^{-1}(n_3/n_2)$], the light will be totally reflected upon striking this interface. The same holds true for a light ray propagating toward the 1-2 interface. The light rays then follow a zig-zag path making the same angle θ_{23} with each reflection. However, because of interference between rays that have traveled different paths, only those rays making certain angles will survive. These are called the modes of the guides. In terms of wave optics, each mode has a different electric field distribution across the guide and propagation constant along the guide. The propagation constant (β) can be determined from wave theory or from the ray-interference pattern. In either case, the wavelength of light in the guide (λ_g) is

$$\lambda_g = 2\pi/\beta$$

and the wave is said to see an equivalent refractive index n_{eq} given by

$$n_{eq} = \lambda_o/\lambda_g$$

where λ_o is the free-space light wavelength. Since the guided wave exists in all three media with n_1, n_2, and n_3, the equivalent index is a sort of average satisfying

$$n_1, n_3 < n_{eq} < n_2 .$$

The electric-field distributions (12) of three modes, TE_0, TE_1, and TE_2 are illustrated in Fig. 2; each has a different n_{eq}. Guides consisting of a $Ga_{1-x}Al_xAs$ layer sandwiched between two layers of $Ga_{1-y}Al_yAs$ of lower index, thus $y > x$ are examples of a one-dimensional guide. Since there are two contacts between layers of dissimilar composition, these are double heterostructure (DH) guides.

Such layered light guiding structures can be readily grown by any of the three previously mentioned crystal growth techniques. Merz and Cho (14) compared optical losses for 1-2 μm thick DH waveguides of similar composition grown by MBE and LPE. They reported losses less than 1.5 cm^{-1} for MBE grown guides between 1.1 and 1.4 eV, the energy of the GaAs DH laser. These losses are comparable to, or less than, the losses observed in similar wave-guide structures grown by the LPE. They did point out, however, that despite the high degree of layer uniformity and control that MBE can achieve, these low losses are not *a priori* obvious. The reason is that the $Ga_{1-x}Al_xAs$ surface exposed during growth in an MBE system is highly reactive which makes the layer very susceptible to contamination by trace amounts of water vapor and carbon-containing gases; these or other contaminants might contribute significantly to the below gap absorption. These results show, however, that is it possible to grow low-loss $Ga_{1-x}AlAs$ wave-guides by MBE with the present-day vacuum technology.

After Gossard *et al.* (9) reported the ultimate in layer control in which up to 10^4 alternating single or multiple monolayers of GaAs and AlAs were sequentially deposited by MBE, Merz *et al.* (15) (see Fig. 3) reported on low-loss waveguides whose central region consisted of alternating sequences of approximately nine monolayers of GaAs and one monolayer of AlAs. The outer layer consisted of an alternating sequence of three monolayers of AlAs and seven monolayers of GaAs. This gave the central region a lower average Al composition ($x \sim 0.10$) than the outer sequence of layers ($x \sim 0.30$). Optical losses were ~1 cm^{-1} higher than the best $Ga_{0.9}Al_{0.1}As$ random alloy waveguides fabricated to date by MBE (14).

Cho, Yariv, and Yeh (16) reported a novel waveguide grown by MBE in which a $Ga_{0.62}Al_{0.38}As$ guiding layer was bounded on one side by air and on the other side by a periodic layered medium composed of alternating layers of GaAs and $Ga_{0.8}Al_{0.2}As$. The periodic media present high refelectivity to radiation which satisfies the Bragg condition. Unlike ordinary dielectric waveguides, confined guiding with arbitrarily low loss is possible even when the guiding layer possesses an index of refraction which is lower than that of the periodic layers.

Various types of three-dimensional heterostructure waveguides have been made by MBE. One attractive approach for making three-dimensional waveguides involves embedding a GaAs waveguide in a matrix of $Ga_{1-x}Al_xAs$. This is achieved by selectively etching away portions of a previously grown planar waveguide resulting in a three-dimensional guide and then epitaxially covering the nonplanar surface with $Ga_{1-x}Al_xAs$ by MBE to completely encapsulate the GaAs waveguide (17). This technique demonstrates one of the growth advantages MBE has over LPE. It is difficult to grow on GaAlAs by LPE once the sample is exposed to the ambient because of the reactive nature of Al. But with MBE the exposed GaAlAs surface need only be cleaned by ion milling prior to any regrowth.

Another technqiue for growing 3-dimensional guides involves etching channels or mesas and growing over them by MBE (18,19). Since the beams can be regarded as being parallel, the local arrival rate is proportional to $\cos \phi$, where ϕ is the local incident angle. When the surface is not flat the incident angle varies point by point on the surface. One of the more interesting guides fabricated by this technique is shown in Fig. 4. It involves the growth in channels aligned along the $(\overline{1}10)$ directions, the two edges of the undercut channels serve as two edge shadowing masks preventing atoms from the incident flux arriving at the surface right underneath them. As a result, a strip of epilayer bounded on all sides by single-crystal planes completely disconnected from the rest of the epilayer is grown in the channel.

CLADDING | CORE

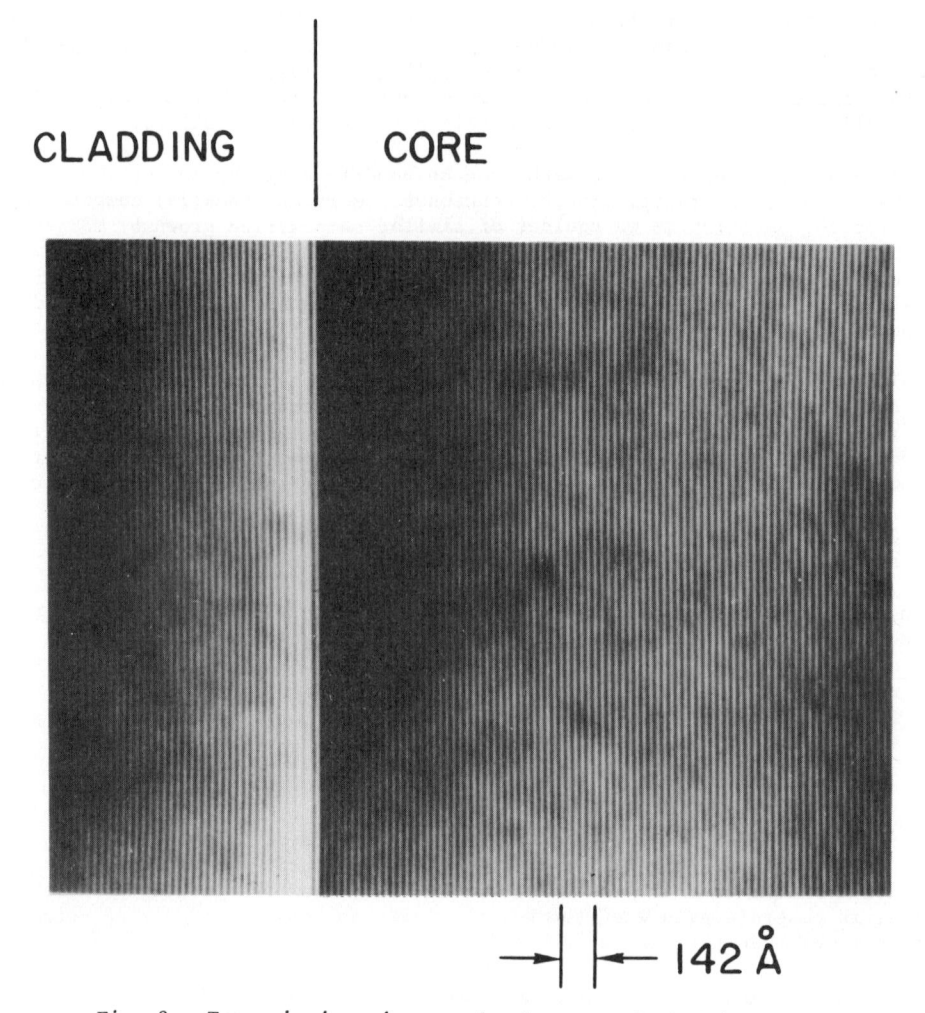

→| |← 142 Å

Fig. 3. Transmission micrograph of core and cladding regions.

Also, because of the directionality of the molecular beam three-dimensional waveguides have been reported by masking molecular beams during epitaxy. As early as 1972 Cho and Reinhart (20) reported on a waveguide structure made by masking with a 50 µm tungsten wire. The Al effusion cell is directed perpendicular to the substrate so that the wire can form a sharp shadow with a width approximately the diameter of the wire. The Ga effusion cell is situated at an angle with respect to the substrate so that the Ga flux does not fall on the same region of the substrate as does the Al flux. This allows the GaAs beam to reach behind the masking wire where the Al beam is shadowed thus resulting in a strip containing only GaAs. Tsang and Ilegems (21) more recently reported on single and multilayered stripe mesa waveguides with widths as narrow as 1 µm using single-level or multilevel masks. These masks were made from single crystal Si wafers. Silicon was chosen as the mask material because it has a low vapor pressure and is relatively inert towards GaAs at the usual growth temperature (~580°C). This makes it possible to press the Si mask directly against the GaAs substrate without contaminating the substrate. The use of single crystalline Si substrates also allows one to generate mask openings with dimensions ≤1 µm by using preferential etchants. Figure 5 shows examples of single and multiple growth through various stripe patterns on single crystal Si masks.

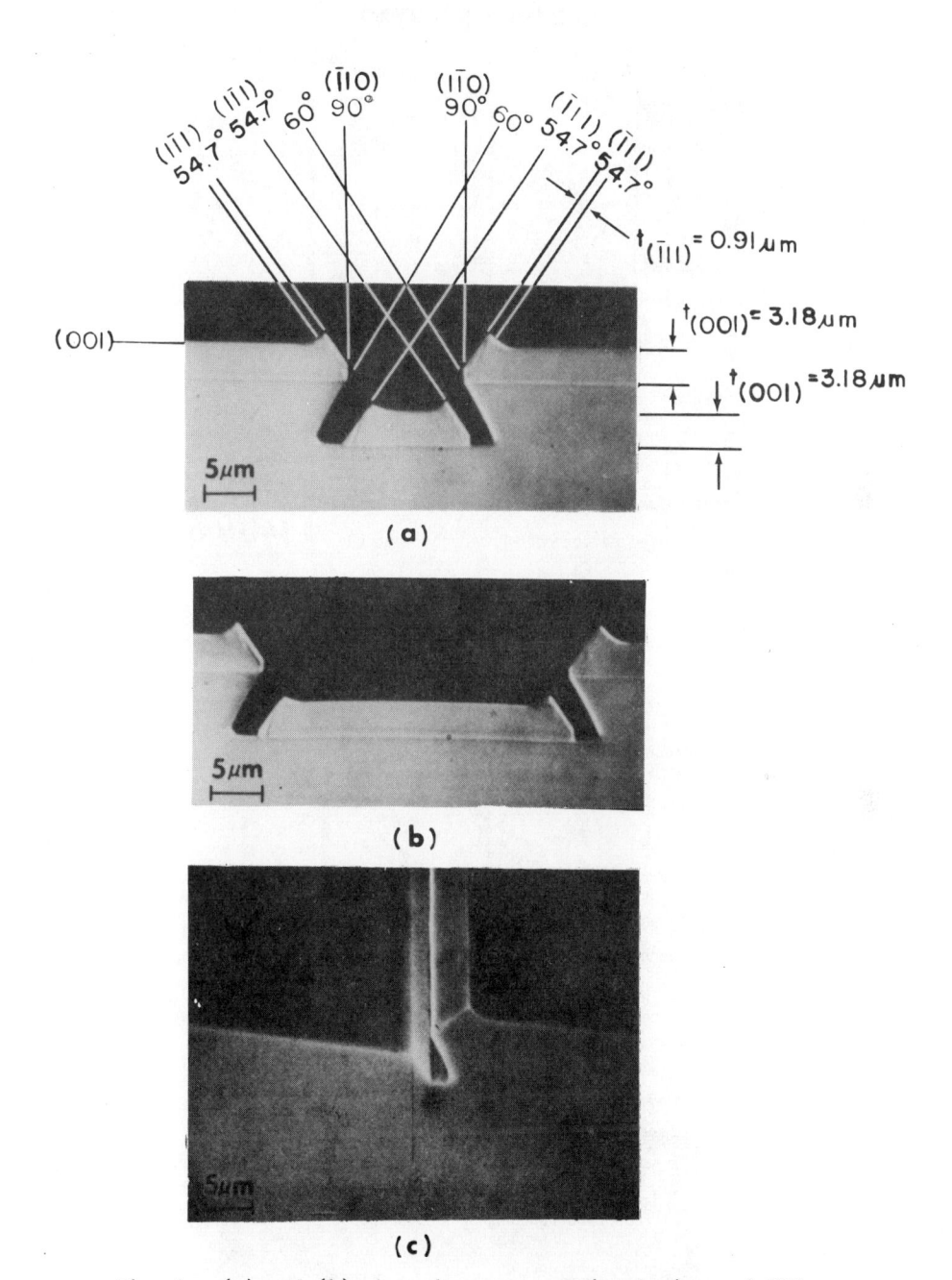

Fig. 4. (a) and (b) show the cross-sectional views of MBE layers over two channels of width 11 and 29 μm aligned along the 110 direction on a (001)-oriented GaAs substrate. (c) A SEM picture of the channel edge showing the surface quality of the as-grown crystal facets.

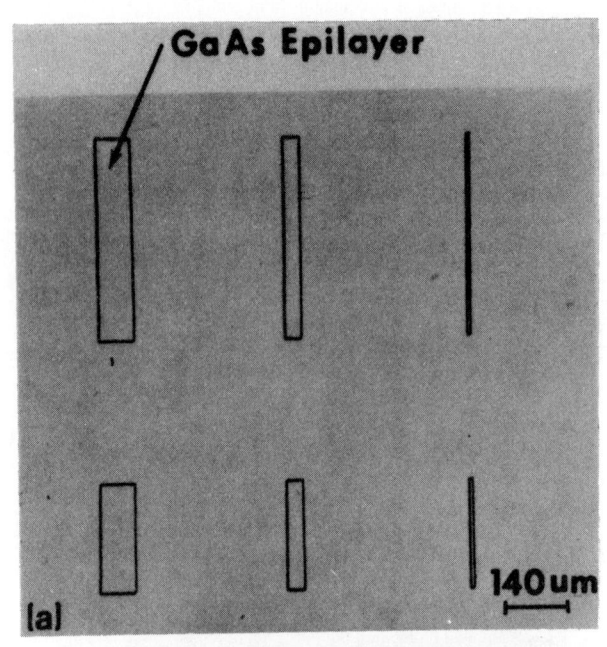

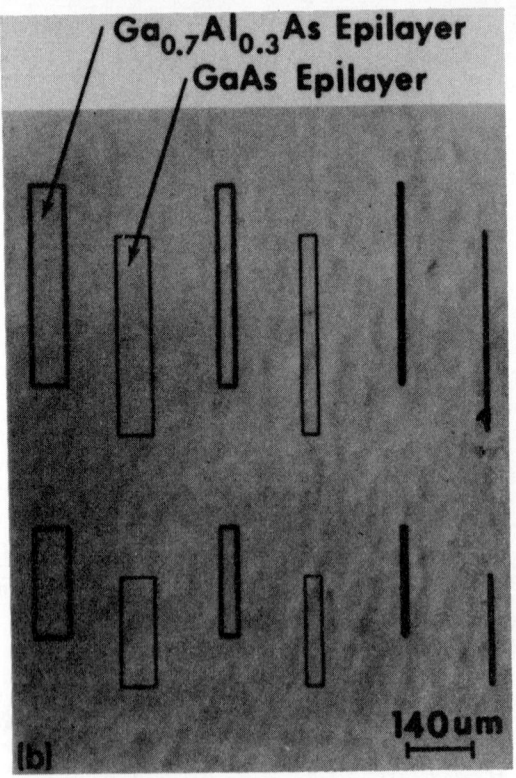

Fig. 5. Optical micrographs of single-level (a) and two-level (b) depositions.

The latest in improvements for making 3-dimensional guides involves molecular beam epitaxial writing of patterned GaAs epilayer structures with controlled thickness profiles, lengthwise varying thicknesses, featureless and optically smooth sidewalls (22). This is achieved by having a fixed Si mask with patterned openings between the evaporating sources and the substrate while the substrate is moved in a plane parallel to the mask during growth (see Fig. 6.).

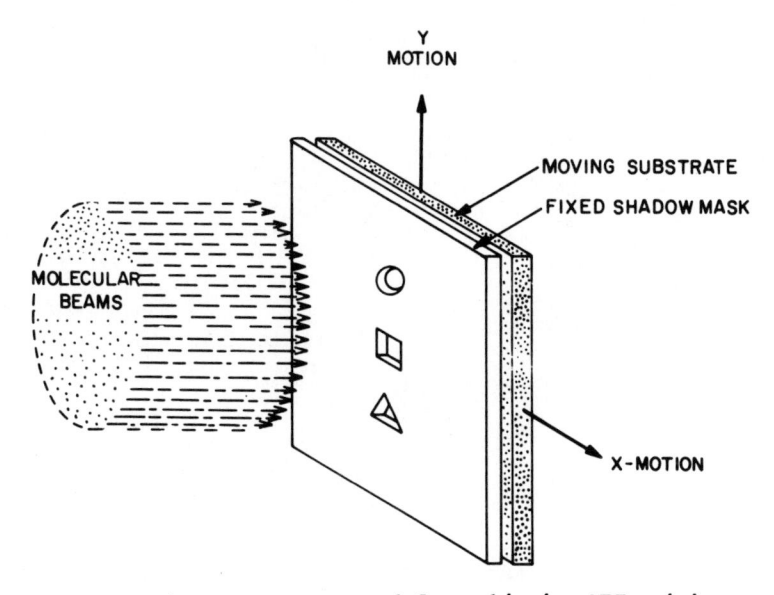

Fig. 6. Basic arrangement used for achieving MBE writing.

GaAs epilayer structures written behind a series of square mask openings with different sizes when the mask-substrate separation is small is shown in Figs. 7a-d.

The various techniques mentioned for fabricating waveguides should give the reader a flavor for the high degree of versatility that is achievable by MBE. It is important to point out that many other types of waveguide structures that have been fabricated by other growth techniques can in principle be fabricated by MBE.

5. DISCRETE LASERS

Ever since the first p-n junction GaAs injection lasers were demonstrated in 1962, considerable effort has been expended in trying to refine the semiconductor laser to the point that it is no longer just a laboratory curiosity but a useful device. Progress was slow until $Ga_{1-x}Al_xAs$ DH lasers were developed in 1970 (23,24). These DH lasers decreased the room temperature threshold current two orders of magnitude over that of initial lasers made only of GaAs. This tremendous reduction in threshold current is attributed to the large refractive index and bandgap difference between the thin $Ga_{1-x}Al_xAs$ (active) layer and the outer two $Ga_{1-y}Al_yAs$ layers when $y - x > 0.15$ and $x \lesssim 0.25$. The large refractive index difference serves to significantly confine the light generated in the $Ga_{1-x}Al_xAs$ (the active region) while the wider bandgap of the $Ga_{1-y}Al_yAs$ causes the injected carriers in the $Ga_{1-x}Al_xAs$ region to be similarly confined. For practical CW lasers it is also necessary to restrict the current to a narrow stripe to limit the active region laterally (25).

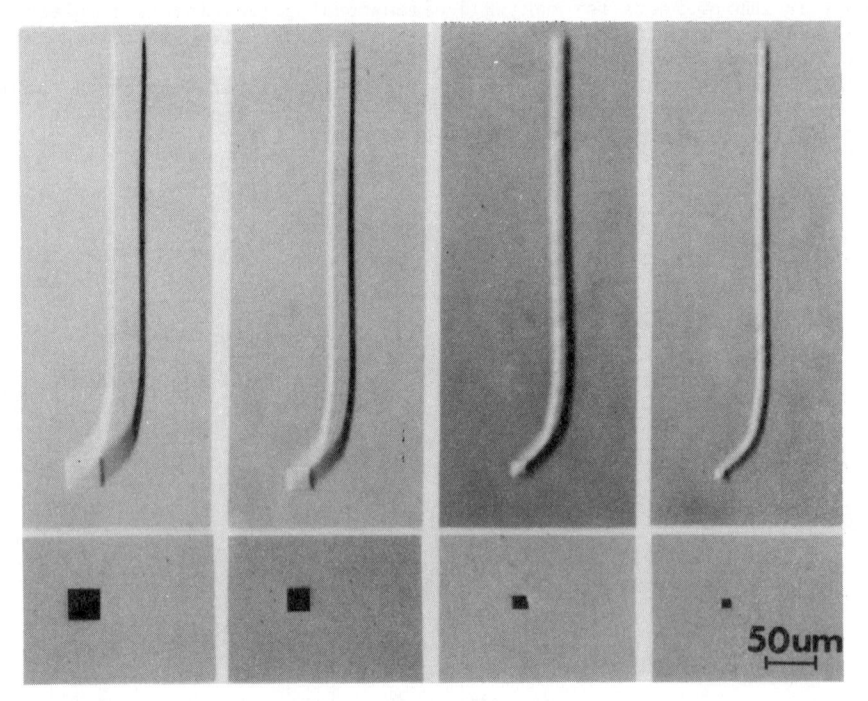

Fig. 7. MBE mask writing through a series of sizes of square mask openings.

Discrete lasers are generally made by cleaving the faces of the semiconductor crystals. These cleaved facets form plane, parallel mirror-like surfaces which reflect a portion of the light back into the region of the p-n junction. The reflected light is amplified, and the energy density within the active waveguide of the laser continues to build up to produce an intense laser beam.

The first GaAs-$Ga_{1-x}Al_xAs$ discrete cleaved facet DH lasers prepared by MBE were reported by Cho and Casey (26). The threshold currents of these lasers were 13 times higher than similar stuctures grown by LPE. It was found that the threshold ratio could be decreased to 1.4 if the samples were annealed at 750°C or above. Contamination of the GaAlAs layers was thought to be the major problem. Later Cho et al. (27), reported CW laser operation at temperatures as high at 100°C. The threshold currents were still about 1.4 times higher but they did not need to be annealed. This was achieved by eliminating more of the hydrocarbon and water vapor sources in the epitaxial growth system by using only pyrolytic BN effusion cells and installing an additional liquid-nitrogen-cooled panel around the susbstrate area. Also, the Al concentrations at all the GaAs-$Ga_{1-x}Al_xAs$ interfaces were graded in order to reduce the lattice mismatch at the interfaces because of the reduced MBE growth temperature 580°C. Another problem with making GaAs and GaAlAs lasers by MBE was to find suitable dopants. Impurities like Zn (28), Cd (28) and Mg (29) have high vapor pressure at the growth temperature compared to the group-III elements. Thus, they tend to have very short absorption lifetimes on the surface and re-evaporate before any significant incorporation in the grown layer can occur. Te (30,31) also has a high vapor pressure at the growth temperature compared to the group-III elements but it has a peculiar problem in that it tends to accumulate on the surface giving nonuniform doping and poor surface morphology. Apparently a stable surface compound (GaTe) is formed. A significant increase in sticking coefficient can be realized by using an ionized dopant beam. Naganuma et al.

(32,33), reported that GaAs layers could be doped heavily p-type $\sim 5 \times 10^{19}$ (cm^{-3}) Zn by ionizing the Zn atoms and weakly accelerating them much like ion implantation. The weakly accelerated Zn ions can penetrate just beneath the growing surface $\sim$10-20 Å, and be incorporated into the grown layer by successive growth. For amphoteric impurities such as the group-IV elements in GaAs, the site distribution ratio appears to be determined (34) by surface stoichiometry during growth. As a result, the elements Si (35), Ge (34,35) and Sn (35) are incorporated primarily as donors under the usual As-rich growth conditions with As-stabilized surface structure, whereas for the case of Ge it has been shown that the same element is incorporated primarily as an acceptor under Ga-stabilized growth conditions. So far, however, it has been difficult to achieve routine mirror-like surface appearance under Ga-stablized growth conditions. Ilegems et al. (36), reported using Mn as a p-type dopant. Even though Mn has a high sticking coefficient it is a $\sim$110 meV deep acceptor and thus doping levels of only 10^{18} cm^{-3} could be achieved. A much more promising impurity is Be (37) where at least 3×10^{19} cm^{-3} Be can be incorporated in GaAs and $Ga_{1-x}Al_xAs$ (up to $x \sim 0.33$). In fact, the Be doping level is simply proportional to the Be arrival rate.

At this point in time Sn appears to be the most attractive n-type dopant and either ionized Zn or Be are attractive for heavy p-type doping. Heavy doping is necessary in order to reduce the series resistance of the device to minimize thermal heating which would allow for CW laser operation.

We have talked about oxygen contamination in GaAlAs as being very deleterious for making DH $Ga_{1-y}Al_yAs/Ga_{1-x}Al_xAs$ lasers. There appears to be a promising aspect to oxygen doped GaAlAs layers and that is they can be made insulating. Casey et al. (38) demonstrated that a lattice matched single crystal insulator-semiconductor heterojunction can be grown by doping $Ga_{0.5}Al_{0.5}As$ layer with oxygen by MBE. This could have significant impact on monolithic device fabrication. For example, this could eliminate the need for an earlier technique developed by Cho and Ballamy (39) for planar technology in which the active device positions are defined by windows in SiO_2 layers which are deposited prior to MBE growth. By this process, polycrystalline and monocrystalline GaAs are selectively grown producing a pattern which consists of areas of device quality single-crystal material isolated by semi-insulating polycrystalline material. In fact, Lee and Cho (40) reported on single transverse mode injection lasers with embedded stripe layers using an SiO_2 mask. A scanning electron-beam micrograph of the cleaved end face of the laser view at an angle is shown in Fig. 8.

Along the same line Cho et al. (41), demonstrated that semiconducting materials consisting of inlays of different doping or composition in selective areas may be grown by MBE. They showed it is possible to fill an etched hole with MBE without formation of voids resulting from facet growth as observed in other growth techniques. Optically pumped laser oscillation from quantum states in thin multilayer heterostructures have been observed (42,43). Energy levels of electrons and holes in a very thin GaAs layer which is sandwiched between wider bandgap layers of $Ga_{0.8}Al_{0.2}As$ consist of a discrete number of bound states within the well. The MBE growth technique not only allows for the formation of rectangular potential wells but by programming the variation of the Al content essentially any arbitrary potential can be repeatedly deposited. Lasers taking advantage of quantum size effects may be able to add a whole new dimension to the improvement of laser properties.

6. LASER INTEGRATION

In order to take full advantage of the concepts of integrated optics it is necessary to fabricate laser structures which can be integrated onto the same substrate as are the other optical and electronic components. Unfortunately, the conventional

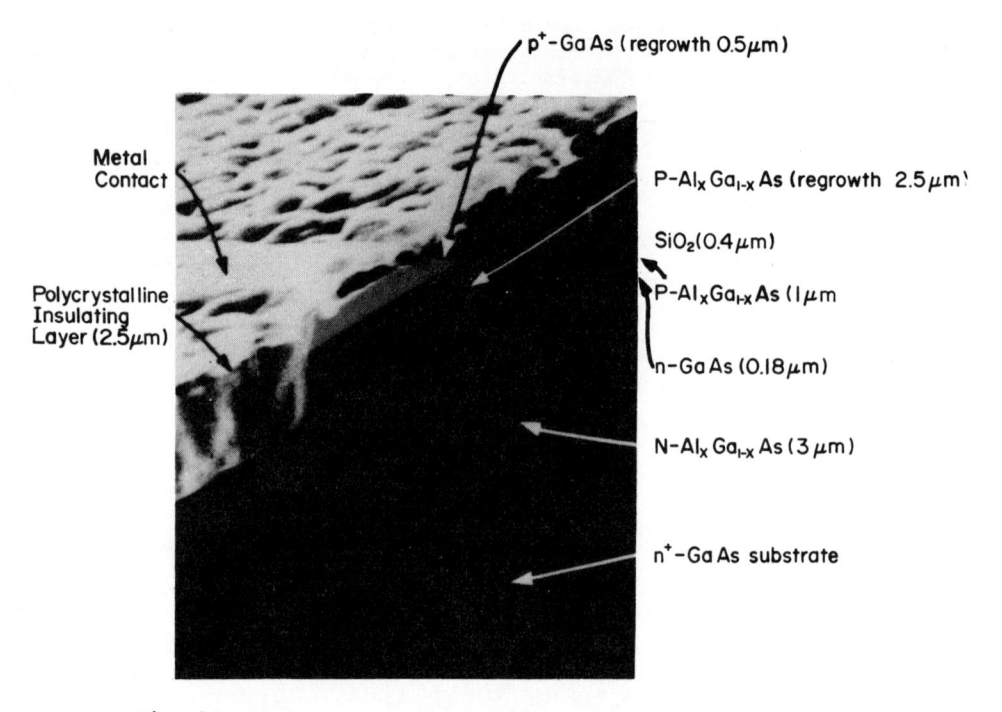

Fig. 8. SEM of the cross-section of an embedded stripe laser.

method of fabricating the mirrors on a semiconductor laser involves cleavage of the crystal. Thus, the laser chip must be broken or cleaved off the substrate in order to provide optical feeback. Within the last several years there have been a number of methods developed for fabricating mirrors or reflecting elements which are compatible with monolithic thin film technology. The first to be realized was distributed feedback (DFB) (44) in which an optical grating is used to diffract incoming light back upon itself. These original devices were grown by LPE. If the period of the grating Λ is an integer multiple of half-wavelengths of light in the guide (λ_g) a small amount of light is reflected back into the waveguide conherently at each corrugation, the sum providing sufficient feedback to operate the laser.

Distributed-feedback heterostructure lasers have several advantages as discrete devices over conventional cleaved-mirror lasers. First, they generate a reproducible laser wavelength, which is four times less sensitive to thermal variation than is a conventional cleaved mirror laser (45,46). Second, the laser usually emits in a narrow-bandwidth single-longitudinal (47,48) mode. Third, as mentioned in part A of this section, the grating in a DFB laser also couples out beams which have approximately 100 times improved directionality (11) over those emitted through the cleaved-end mirror or a conventional laser diode. Thus, these DFB devices, in addition to serving as sources for integrated optical circuits, may also prove useful as stand-alone discrete devices. One problem that does exist with DFB lasers is that non-radiative recombination centers are introduced by the grating. One solution is to physically separate the corrugations from the active region (11). Casey et al. (49), found that ion milling of the corrugations did not affect the radiative recombination as long as the corrugations were removed 0.30 µm or more from the layer confining the carriers. This involved a hybrid LPE and MBE growth process (see Fig. 9). MBE is used to grow over the corrugations since growth on a GaAlAs surface is not reproducible using LPE.

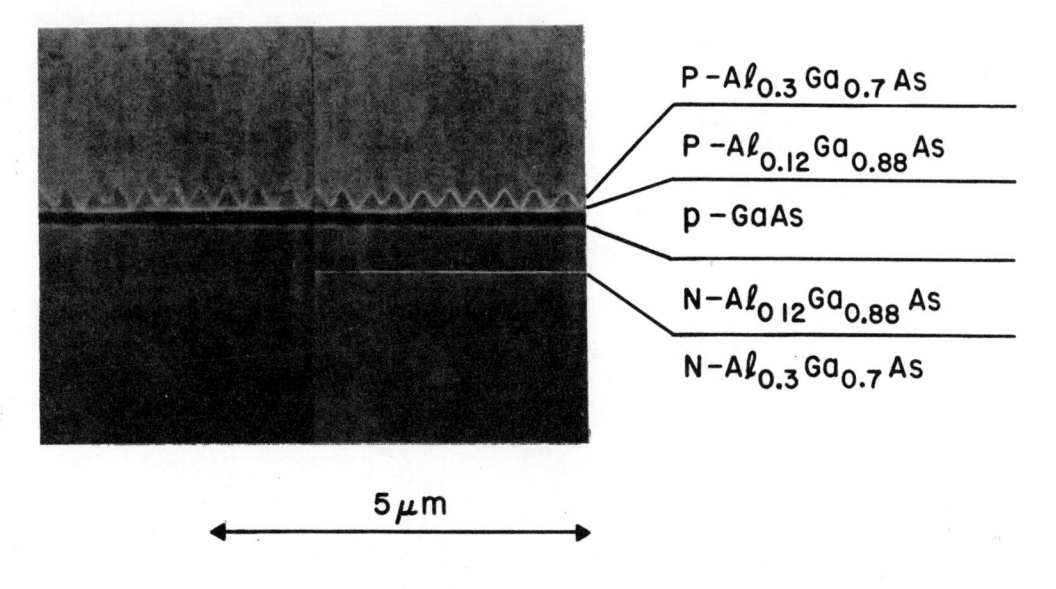

Fig. 9. SEM cross-section of a DFB laser.

It was suggested by S. Wang (50) that instead of gratings being used to provide distributed reflection throughout the laser cavity, they could be put at the ends of the cavity to act as mirrors. Lasers of this type are called DBR (Distributed Bragg Reflection) lasers, and they appear to have more promise than DFB lasers. The removal of the grating from the active region has the obvious advantage of eliminating the nonradiative recombination associated with it, thus allowing simpler structures and simpler processing for the DBR laser. The excellent layer thickness and composition control afforded by MBE should prove quite valuable in fabricating DBR lasers since such control makes it possible to determine a priori the correct grating spacing. The DBR laser should also have at least as good frequency and mode control as the DFB laer.

Integration of DH lasers by fabricating discrete parallel reflectors without cleaving the crystal has also been achieved by selective ion sputtering (51), chemical etching (52-53) and epitaxy (54). Although these devices were not grown by MBE, they could have easily been with the possible exception of Ref. 54 since it involves facet formation along with selective epitaxy.

7. COUPLERS

As mentioned earlier the interconnection of optical components via waveguides is the central problem of monolithic integrated optics and is at the mercy of the crysal grower and the growth technique he chooses to use. MBE appears to be particularly suitable for making structures which allow coupling from one waveguide to another. Conwell and Burnham (11) discuss essentially five kinds of couplers which are: prism, grating, butt and compositional, tapered and directional or evanescent. Prism, grating, and butt couplers are mainly used to couple light out of the optical circuit and will not be discussed in this paper. The tapered coupler has received considerable attention for coupling between waveguides. A tapered coupler is made by tapering down the end of a waveguide. When a guide mode that has been undergoing successive reflections with the angle θ_m reaches the tapered region its reflections take place at progressively smaller θ_m because of the shape of the taper, until θ_m becomes smaller than the critical angle. Beyond that point it is no longer confined by the guide, and is refracted into an adjacent waveguide of slightly lower index of refraction. Taper couplers are quite easily made by MBE and are extremely uniform, gradual and reproducible (see Fig. 10). Prior to starting a growth which is to have a taper an 0.25 mm thick knife-edged mask is swung into place 1 or 2 mm above the surface of the sample. This masks a portion of the substrate from the Ga and As sources. A GaAs layer grows on the unmasked regions, terminated by linear tapers ~200 μm wide which grew in the penumbra areas of the mask edge. Merz and coworkers report (55) that coupling efficiencies approaching 100% can be routinely achieved by MBE while efficiencies of only ~50% have been achieved by LPE.

Composition coupling involves the transition from one composition to another along a waveguide. Various types of composition coupling schemes have been reported, however, none involves MBE. With the larger repertoire of MBE growth techniques there should be no problem in making composition couplers by MBE. On the other hand, with taper couplers approaching 100% efficiency there does not seem to be much incentive to develop composition couplers.

When two identical guides are sufficiently close (within about a wavelength of light) that the exponentially decaying (evanescent) light from one extends into the other the energy will oscillate between the guides with a characteristic period L_o called the coupling lenth. Evanescent wave coupling has been used in twin guide lasers (51). Here the active waveguide of the laser is separated by a region of lower index from a passive guide separation being small enough so that the evanescent fields overlap. Twin guide lasers so far have only been fabricated by LPE but it is obvious that MBE is much more suitable for its fabrication since layer thicknesses and index of refraction can be controlled so readily.

8. SWITCHES, MODULATORS AND DETECTORS

We refer the reader again to the review article by Conwell and Burnham (11) for a more complete coverage of switches, modulators and detectors related to GaAs and its alloys. It suffices to say that although none of these components have been fabricated by MBE techniques these principles are well understood and have been fabricated by other growth techniques and MBE does not seem to present any limitations.

9. MULTI-COMPONENT INTEGRATION

Although there has been various types of multi-component integration including the integration of a laser, waveguide and detector recently reported by Merz and Logan (56), only one involves growth by MBE. Reinhart and Cho (57) reported growth of a $Ga_{1-y}Al_yAs$-$Ga_{1-x}Al_xAs$ laser taper coupled to a passive (more transparent) waveguide grown by MBE. The structure is similar to the schematic shown in Fig. 10a except additional epitaxial layers are grown to facilitate electrical contact. Layer t_o is the active layer of the devices and layer $t_{0.1}$ to the right of the taper services as the passive waveguide. These particular structures indicated significant absorption losses.

10. CONCLUSIONS

MBE has certainly demonstrated that monolayer dimensional control along with a wide range of doping and compositional profiles is achievable. Because of this versatility, MBE is a promising growth technique for the fabrication of extremely compli-

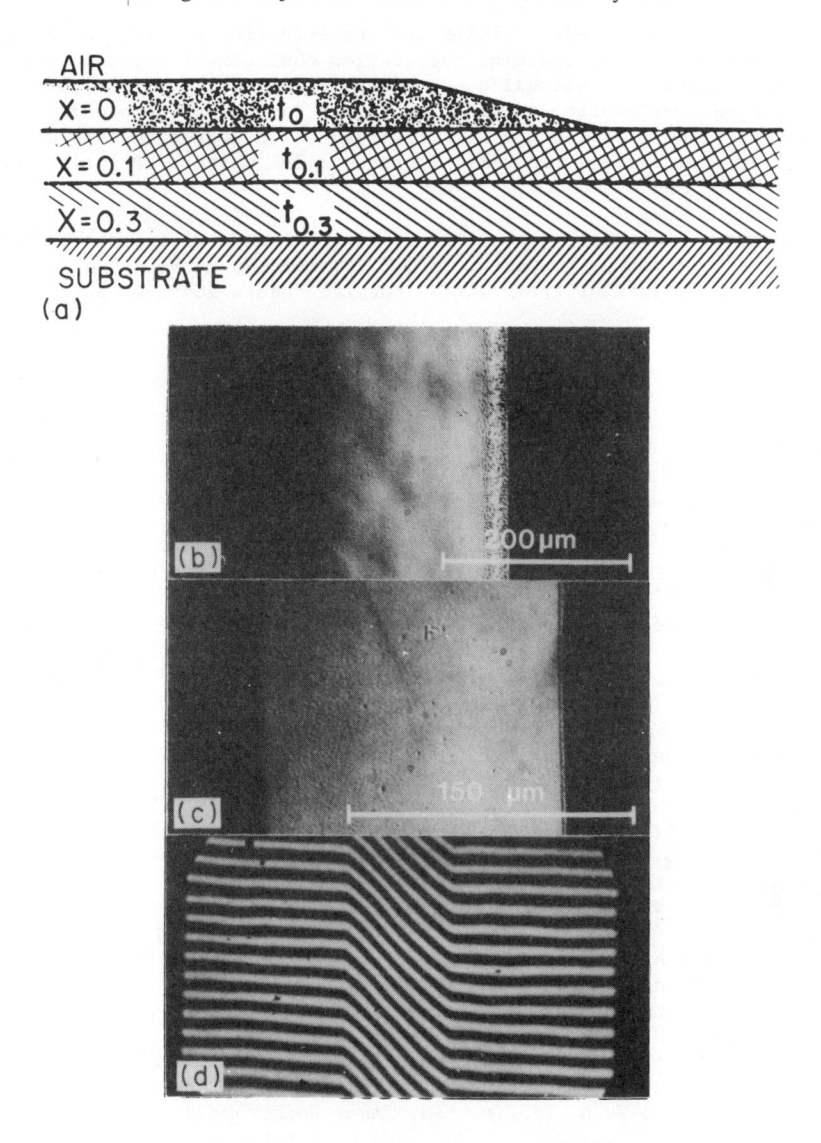

Fig. 10. Tapers grown by MBE.

cated integrated optical circuits. However, considerable effort and improvement is needed to identify and suppress the sources of contamination during the growth of GaAs and especially GaAlAs. Already improved system designs are beginning to be implemented such as the incorporation of elaborate vacuum interlocks, improved cryopumping, improved substrate heaters and source heater assemblies. These steps should significantly reduce comtamination during growth. Calawa (58) recently reported that the introduction of hydrogen during MBE growth of GaAs seems to significantly improve the mobility of the material. As soon as MBE grown GaAlAs can reproducibly compare with LPE grown GaAlAs, then significant advances in integrated optics can be expected. Solving the contamination problem in MBE is sort of analogous to solving the waveguiding and carrier confinement problems for lasers. As was mentioned earlier, this was achieved by making DH lasers. Without it laser progress was slow. Of course, solving the GaAlAs contamination problem may, on the other hand, simply require the use of alloys which do not contain Al. For example, Miller *et al.* (59), recently reported on InP/GaInAs/InP lattice matched DH lasers grown by MBE.

We are grateful to R. Z. Bachrach and A. Y. Cho for many helpful discussions. Acknowledgements are due H. C. Casey, Jr., T. P. Lee, J. L. Merz and W. T. Tsang for supplying photographs of their work. Special thanks go to D. Bernsen for secretarial service.

REFERENCES

1. S. E. Miller, *Bell Syst. Tech. J.* **48**, 2059 (1969).
2. R. D. Dupuis and P. D. Dapkus, *Appl. Phys. Lett.* **32**, 406 (1978).
3. A. Y. Cho and F. K. Reinhart, *J. Appl. Phys.* **45**, 1812 (1974).
4. J. R. Arthur and J. J. LePore, *J. Vacuum Sci. Technol.* **6**, 545 (1969).
5. A. Y. Cho and M. B. Panish, *J. Appl. Phys.* **43**, 5118 (1972).
6. L. L. Chang, L. Esaki, W. E. Howard, R. Ludeke, and G. Schul, *J. Vacuum Sci. Technol.* **10**, 655 (1973).
7. R. Dingle, W. Wiegmann, and C. H. Henry, *Phys. Rev. Lett.* **33**, 827 (1974).
8. R. Dingle, A. C. Gossard, and W. Wiegmann, *Phys. Rev. Lett.* **34**, 1327 (1975).
9. A. C. Gossard, P. M. Petroff, W. Wiegmann, R. Dingle, and A. Savage, *Appl. Phys. Lett.* **29**, 323 (1976).
10. P. K. Tien, *Rev. Mod. Phys.* **49**, 361 (1977).
11. E. M. Conwell and R. D. Burnham, *Ann. Rev. Mater. Sci.* **8**, 135 (1978).
12. M. K. Barnoski, ed., *Introduction to Integrated Optics*, Plenum, New York (1974).
13. T. Tamir, ed., *Integrated Optics*, Springer, New York (1975).
14. J. L. Merz and A. Y. Cho, *Appl. Phys. Lett.* **28**, 456 (1976).
15. J. L. Merz, A. C. Gossard, and W. Wiegmann, *Appl. Phys. Lett.* **30**, 629 (1977).
16. A. Y. Cho, A. Yariv, and P. Yeh, *Appl. Phys. Lett.* **30**, 471 (1977).
17. J. C. Tracy, W. Wiegmann, R. A. Logan, and F. K. Reinhart, *Appl. Phys. Lett.* **22**, 511 (1973).
18. W. T. Tsang and A. Y. Cho, *Appl. Phys. Lett.* **30**, 293 (1977).
19. S. Nagata, T. Tanaka, and M. Fukai, *Appl. Phys. Lett.* **30**, 503 (1977).
20. A. Y. Cho and F. K. Reinhart, *Appl. Phys. Lett.* **21**, 355 (1972).
21. W. T. Tsang and M. Illegems, *Appl. Phys. Lett.* **31**, 301 (1977).
22. W. T. Tsang and A. Y. Cho, *Appl. Phys. Lett.* **32**, 491 (1978).
23. Zh. I. Alferov, V. M. Andreev, D. Z. Garbuzov, Yu. V. Zhilgaev, E. P. Morozov, E. L. Portnoi, and V. G. Trofim, *Fiz. Tekh. Poluprovodn* **4**, 1826 (1970).
24. I. Hayashi, M. B. Panish, P. W. Foy, and S. Sumski, *Appl. Phys. Lett.* **17**, 109 (1970).
25. J. E. Ripper, J. C. Dyment, L. A. D'Asaro, and T. L. Paoli, *Appl. Phys. Lett.* **18**, 155 (1971).
26. A. Y. Cho and H. C. Casey, Jr., *Appl. Phys. Lett.* **25**, 288 (1974).
27. A. Y. Cho, R. W. Dixon, H. C. Casey, Jr., and R. L. Hartman, *Appl. Phys. Lett.* **28**, 501 (1976).
28. J. R. Arthur, *Surf. Sci.* **38**, 394 (1973).
29. A. Y. Cho and M. B. Panish, *J. Appl. Phys.* 43, 5118 (1972).
30. J. R. Arthur, *Surf. Sci.* **43**, 449 (1974).
31. A. Y. Cho and J. R. Arthur, *Prog. Solid. St. Chem.* **10**, 157 (1975).
32. M. Naganuma and K. Takahashi, *Appl. Phys. Lett.* **27**, 342 (1975).
33. N. Matsunaga, M. Naganuma, and K. Takahashi, *Jap. J. Appl. Phys.* **16**, 443 (1976).
34. A. Y. Cho and I. Hayashi, *J. Appl. Phys.* **42**, 4422 (1971).
35. A. Y. Cho, *J. Appl. Phys.* **46**, 1732 (1975).

36. M. Ilegems, R. Dingle, and L. W. Rupp, Jr., *J. Appl. Phys.* 46, 3059 (1975).
37. M. Ilegems, *J. Appl. Phys.* 48, 1278 (1977).
38. H. C. Casey, Jr., A. Y. Cho, and E. H. Nicollian, *Appl. Phys. Lett.* 32, 678 (1978).
39. A. Y. Cho and W. C. Ballamy, *J. Appl. Phys.* 46, 783 (1975).
40. T. P. Lee and A. Y. Cho, *Appl. Phys. Lett.* 29, 164 (1976).
41. A. Y. Cho, J. V. Dilorenzo, and G. E. Mahoney, *IEEE Trans. on Electr. Devices* ED-24, 1186 (1977).
42. J. P. van der Ziel, R. Dingle, R. C. Miller, W. Wiegmann, and W. A. Nordland, Jr., *Appl. Phys. Lett.* 26, 463 (1975).
43. R. C. Miller, R. Dingle, A. C. Gossard, R. A. Logan, W. A. Nordland, Jr., and W. Wiegmann, *J. Appl. Phys.* 47, 4509 (1976).
44. D. R. Scifres, R. D. Burnham, and W. Streifer, *Appl. Phys. Lett.* 25, 203 (1974).
45. M. Nakamura, K. Aiki, J. Umeda, A. Katzir, A. Yariv, and H. W. Yen, *IEEE J. Quantum Electr.* QE-11, 436 (1975).
46. R. D. Burnham, D. R. Scifres, and W. Streifer, *Appl. Phys. Lett.* 29, 287 (1976).
47. R. D. Burnham, D. R. Scifres, and W. Streifer, *IEEE J. Quantum Electr.* QE-11, 439 (1975).
48. D. R. Scifres, R. D. Burnham, and W. Streifer, *IEEE Trans. on Electr. Devices* ED-22, 609 (1975).
49. M. Ilegems, H. C. Casey, S. Somekh, and M. B. Panish, *J. Crystal Growth* 31, 158 (1975).
50. S. Wang, *IEEE J. Quantum Electr.* QE-10, 413 (1974)
51. Y. Suematsu, M. Yamada, and K. Hayashi, *IEEE J. Quantum Electra.* QE-11, 457 (1975).
52. C. Hurwitz, J. A. Rossi, J. J. Hsieh, C. M. Wolfe, *Appl. Phys. Lett.* 27, 241 (1975).
53. J. L. Merz and R. A. Logan, *Appl. Phys. Lett.* 47, 3503 (1976).
54. J. C. Campbell and D. W. Bellavance, *IEEE J. Quantum Electr.* QE-13, 253 (1977).
55. J. L. Merz, R. A. Logan, W. Wiegmann, and A. C. Gossard, *Appl. Phys. Lett.* 26, 337 (1975).
56. J. L. Merz and R. A. Logan, *Appl. Phys. Lett.* 30, 530 (1977).
57. F. K. Reinhart and A. Y. Cho, *Appl. Phys. Lett.* 31, 457 (1977).
58. A. R. Calawa, 20th Annual Electronic Materials Conference, Santa Barbara, California, June 28-30, 1978.
59. B. I. Miller, J. H. McFee, R. J. Martin, and P. K. Tien, *Appl. Phys. Lett.* 33, 44 (1978).

THE AUTHORS

R. D. Burnham

D. R. Scifres

Robert D. Burnham was born in Havre de Grace, Maryland, on 21 March 1944. He received his B.S., M.S., and Ph.D. degrees from the University of Illinois, Urbana, in 1966, 1968, and 1971, respectively. He held an NDEA Fellowship from 1966-1969, and a General Telephone and Electronics Fellowship from 1969-1971.

He has been a member of the Research Staff at the Xerox Palo Alto Research Center, Palo Alto, California, since 1971. He is currently in charge of the materials growth of III-V semiconductor lasers by molecular beam epitaxy, liquid phase epitaxy and metalorganic chemical vapor deposition.

He has been a co-author of over eighty papers. His most significant contributions were in the area of distributed feedback, quaternary, buried heterostructure, and InGaP lasers.

Dr. Burnham is a member of Tau Beta Pi, Eta Kappa Nu, Sigma Tau, the Electrochemical Society, and the AIME.

Don R. Scifres was born in Lafayette, Indiana, on 10 September 1946. He received the B.S. degree with honors in Electrical Engineering from Purdue University, West Lafayette, Indiana, in 1968, and the M.S. and Ph.D. degrees in Electrical Engineering from the University of Illinois, Urbana, Illinois, in 1970 and 1972, respectively. While at the University of Illinois he held a University of Illinois Doctoral Support Fellowship and a General Telephone and Electronics Fellowship.

In 1972, he joined the Xerox Palo Alto Research Center where he is currently Manager of an Opto-Electronics Group. His research has been involved with developing semiconductor injection lasers as well as incorporating optical and electronic components in the semiconductor laser chip to achieve such integrated optical functions as distributed feedback, transverse mode control, and electronically controlled laser beam deflection.

Dr. Scifres is on the Board of Editors for the journal of *Fiber and Integrated Optics*, is a national lecturer on "Semiconductor Lasers" for the Quantum Electronics group of IEEE, is a Fellow of the Optical Society of America, and a member of IEEE and the American Physical Society.

SEMICONDUCTOR SURFACE AND CRYSTAL PHYSICS STUDIED BY MBE

R. Z. Bachrach

XEROX PARC, 3333 Coyote Hill Road, Palo Alto, CA 94304, U.S.A.

(Submitted August 1978)

Abstract

A review is presented of semiconductor surface and crystal physics studies carried out with the use of molecular beam epitaxial growth techniques. The studies reviewed are related to an understanding of the various aspects of the growth process. Examples are drawn from a variety of semiconductors, but special emphasis is given to GaAs related results.

1. Introduction

Molecular Beam Epitaxy[1,2,3] has evolved continuously from relatively uncontrolled free evaporation in low vacuum to its present state today where ultra-high vacuum apparatus is used and significant control can be exercised. This evolution has been driven by the desire to use evaporative techniques to grow "device quality" crystals. Today in fact, as discussed elsewhere in this issue, it is possible to grow a wide variety of semiconductor crystals in device structures as lasers, diodes or microwave transistors which are competitive with devices fabricated from material grown by other methods. Underpining the evolution of evaporative methods into MBE has been the use of *in-situ* diagnostic techniques which have allowed the epitaxy process to be decomposed and problems isolated. Thus as MBE has evolved, the application of surface analysis tools have had an instrumental impact on the success of the evolution.

This article will review a variety of surface and bulk studies which have provided new information on semiconductor surfaces and which have clarified the physical processes involved in molecular beam epitaxial growth. These studies have been both *in-situ* during and after growth and with separate experiments on the crystals or structures subsequent to growth. These latter experiments are often interesting because they are, in fact, dependent on the structures that MBE can grow. As MBE has been better understood it is in turn being used for surface studies which previously had been restricted to cleaved samples.

The results and techniques that will be discussed have typically evolved concurrently with their first use in MBE experiments. Similarly the application of new vacuum pumping technologies has run parallel to their introduction. These hardware aspects are discussed in the chapter by Luscher and Collins.[4] It is both accessible and commonplace to configure stainless steel MBE systems[5] capable of 5×10^{-12} - 5×10^{-11} Torr base pressure with a multiplicity of electron, ion and photon spectroscopies.[6,7]

MBE is often referred to as a highly instrumented technique. This has occurred in the evolution from the technique of free evaporative deposition because of the complexity of the overall process. The initial failures to achieve deposition of semiconductor material with the desired qualities required that experiments be performed to understand the overall process. This included determination of evaporative species, adsorption processes and resulting surface structures at the growth temperature.

Three techniques contributed significantly to the progress of MBE. These are quadropole residual gas analysis, high energy electron diffraction analysis,[8,9] and Auger electron spectroscopy.[5,10] Most research systems now include some aspects of these techniques.

The utilization of quadropole mass analyzers allowed the effective diagnosis of the gas ambient as well as the investigation of source composition and desorption phenomena. Auger spectroscopy is very powerful for examining surface cleanliness and composition. High energy electron diffraction, which was developed in the 1930's, has had an important impact on the development of epitaxy since it allows crystallinity to be determined at growth initiation. HEED and then LEED in turn have shown the complexity of surface structures that exist on semiconductor surfaces and in co-ordination with Auger the dependence of the structures on composition.

An understanding of the growth ambient and its interaction with a cleaned substrate is an important aspect of MBE. The contamination resulting from the background molecular adsorption is in itself an interesting area of study. The use of UHV for MBE resulted from an attempt to minimize the contamination resulting from unintentional adsorption processes. The sticking coefficient for a particular species relative to the growth rate is an important parameter determining imparity incorporation in most cases, adsorption to the metals before they have reacted is more important than interaction with the semiconductor formed. The contaminants can be from several sources. The ambient gas background, reacted gases on the sources, other hot filaments or unintentional evaporants from the sources. The result of these effects can range from disruption of epitaxy, to poor control of doping, to poor control of minority carrier lifetime. Aspects of these surface and bulk related problems are described in this review.

A fairly good understanding of crystal growth with MBE exists today, however many parameters and aspects remain to be explored more deeply. For example, the achievable growth rates as a function of crystal "quality" are not established. Recent evidence would indicate that the optimum growth temperatures are also not firmly established. Many of the questions that remain relate to the interplay of surface structures with doping phenomena. Interdiffusional processes are also receiving more attention. These latter questions will directly impact on the types of devices and their dimensions that can be grown with MBE.

2. Substrate preparation

The epitaxial process typically starts with a crystalline substrate inserted from room ambient after some preparation. The substrate typically is covered with a contamination layer consisting of carbonaceous compounds and oxides.[2,11] Two examples for discussion are GaAs and Si. In both these cases, the final polishing preparation is often terminated in such a way that an oxide film is left. Surface studies have clarified the preparation necessary to initiate epitaxial growth.

In the case of GaAs, several different cleaning regimes exist which depend on how well carbon is excluded from the GaAs-ioxide interface upon termination of polishing. In the worst case for GaAs heat treatment in vacuum can remove the oxide, but ion sputtering is required to remove the carbonaceous component. This is exemplified in Fig. 1 which shows Auger derivative spectra at three stages of cleaning. Oxygen desorption becomes appreciable above 500°C, and the oxide film is removed within several minutes. These cleaning procedures leave the surface potentially in a variety of surface reconstructions which are discussed in the next section. Studies have shown that contamination must be removed to 0.1 monolayer or less in order for the epitaxy to proceed well. Once a clean surface is achieved, annealing can restore crystallinity. Typically, however, the resulting surface will not be stoichiometric.

If the surface is prepared properly in terminating the polishing, the need for Argon sputtering can be avoided. The heat treatment required is as high as 650° to remove all the carbon and oxygen. Stoichiometry is restored with the initial arsenic flux prior to initiating growth.

Similar procedures work for silicon, but silicon is more reactive. The silicon surface is particularly sensitive to carbon contamination, presumably through the formation of silicon carbide. The carbonaceous contamination can disrupt epitaxy. Si MBE which produces "device quality" layers seems to be best achieved in helium cryopumped vacuum systems. Such systems achieve lower partial pressures of carbon containing molecules such as CO, CO_2, CH_4 etc. Silicon cleaning is typically more difficult and temperatures in the 1200°C range are used. Argon ion sputtering can remove these layers, but damage is introduced which must be annealed prior to

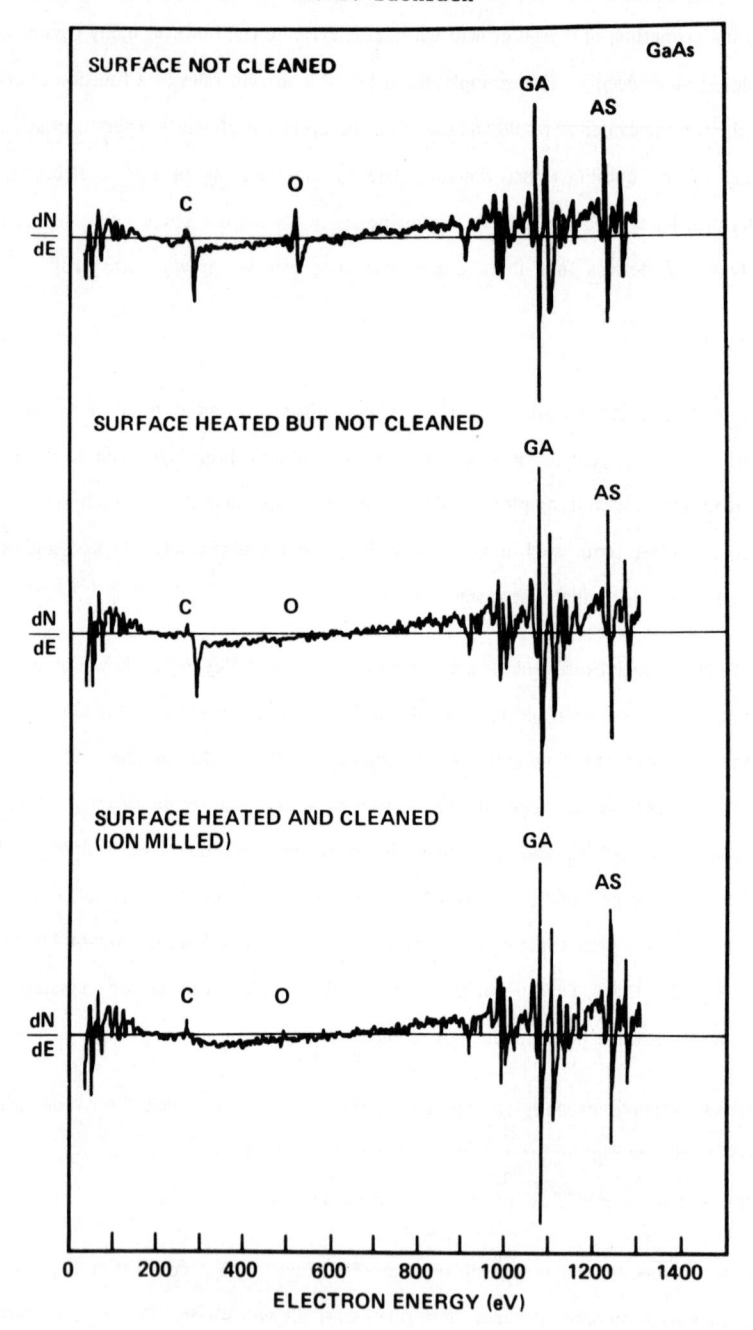

Fig. 1 Auger spectra showing the cleaning stages of GaAs. The upper spectrum shows the surface after polishing and insertion from ambient. The middle spectrum shows the surface after heating to 550°C which removes the oxides. Removal of the carbon shown in the bottom spectrum requires argon ion bombardment.

initiation of epitaxy. Thermal cleaning appears to be the preferred method for silicon to date. Bean, *et al.* have investigated the damage associated with argon sputtering using He backscattering.[12] They find that although the surface damage can be annealed away for a wide variety of conditions as observed with *in situ* techniques such as HEED and AES, other techniques which are more sensitive show that the extent and concentrations of defects depend strongly on substrate temperature. Cleaning with a cold substrate and then annealing leaves the best substrate from which to initiate epitaxy.

3. Surface structures

Surface structures of semiconductors have been under active investigation for many years.[13] The free surfaces tend to undergo both structural reconstruction and relaxation. These effects depend on the surface composition in the case of GaAs or other III-V's. Overlayers add further complication.

LEED patterns can provide surface structures in the vicinity of room temperature. HEED is able to provide information at the growth temperature.[14] The information provided by this technique has been crucial in bringing MBE under control, by providing an active monitor.

The importance of surface structures will appear repeatedly in the following discussions of effects related to surface states, doping effects and overlayers. At the outset therefore the structural aspects have to be illucidated. Many of these surface structures represent phases with varying surface composition while some represent reconstructions without compositional change. In addition to the structure, positional relaxations can occur which maintain the bulk periodicity. These effects can be exemplified by discussion of the epitaxy of Si, GaAs and ZnSe or from covalent to relatively ionic crystals.

Fig. 2 shows schematically the side and top view of the zincblende structure (100), (110) and (111) surface. These low index faces are typically used for epitaxy. The (100) surface is used for GaAs laser related work because of two parallel (110) planes can be used to cleave mirror faces. Examining these figures is instructive since one can readily see the origin of the reconstructions that occur and the plausible way they are related to stoichiometry.

The complexity of these surface structures is exemplified by a study of the MBE of ZnSe on GaAs.[15] These two semiconductors nominally lattice match. However in the growth on a GaAs (100) surface, ZnSe forms in a C (2x2) and even for a thick film retains this surface structure. Silicon (100) typically exhibits a 2x1 surface reconstruction, and the (111) face is quite complex.[16]

ZINCBLENDE

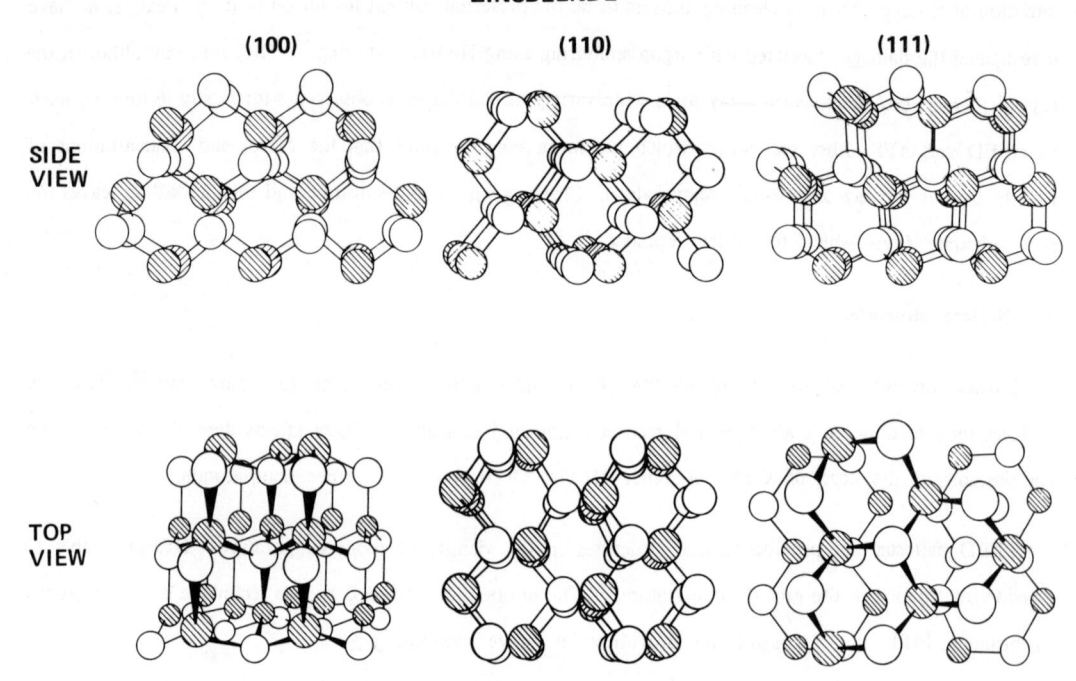

Fig. 2 Zincblende (100), (110) and (111) ideal surface models showing side and top view. The (110) surface exhibits a (1x1) reconstruction where the As moves out of the plane and the Ga into the plane. The top view shown is translated with respect to the side view. The (100) and (111) surface exhibit a variety of stoichiometry related reconstructions discussed in the text. (fig. after C. B. Duke, ref. 13)

The following discussion will focus on GaAs since much of the work has involved this III-V semiconductor. Fig. 2 represents the ideal surface termination. Referring to the side view, the distance between the outermost layers are $1/4 a_c$, $1/[2(2^{1/2})]a_c$, and $(3^{1/2}/4)a_c$ respectively for the (100), (110), (111) faces. For GaAs at room temperature, $a_c = 5.657 Å$. The (110) face is now known to relax in such a way that the As moves out by .2Å and the Ga moves in by 0.45Å.[17] The displacement of 0.65Å is comparable to the 0.82Å interplanar separation for the ideal (111) face. All of the real faces then display a layer of atoms where one or the other species is predominantly exposed. In the case of the (100) and (111) faces, this leads to a variety of reconstructions which are determined by the site occupancy.

Studies of GaAs by Cho showed that two principal regimes occur on the (100)[18] and (111)[19] faces and the transition between them is driven by the surface stoichiometry. This result was established conclusively in a study by Arthur in which structural determination was co-ordinated with Auger.[20] The details of these surface structures have been extended by a number of investigators.[21-23] The issue arises since stoichiometric changes in surface position can also result in reconstructions. This in fact, happens on the Si surface.

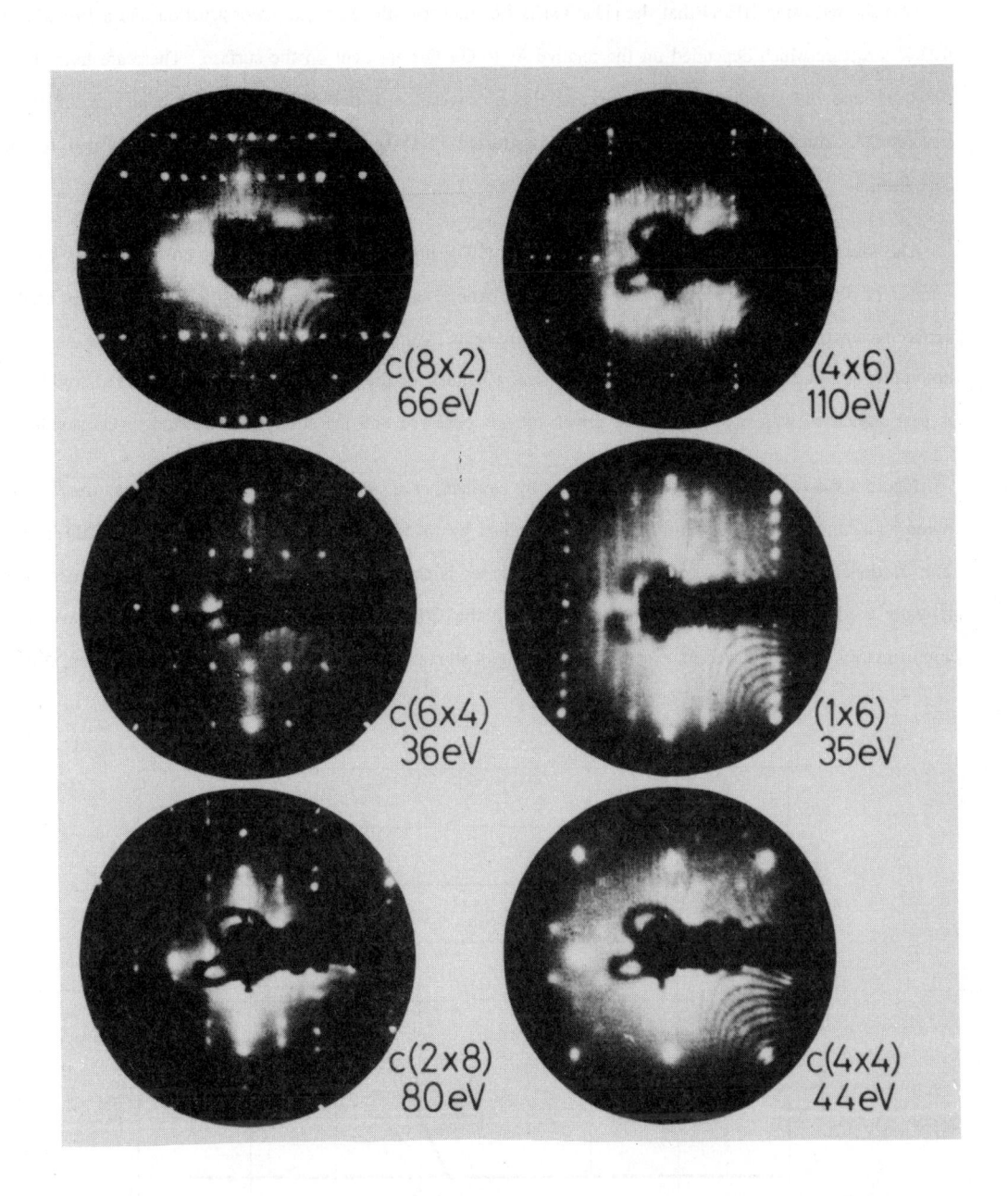

Fig. 3 LEED photographs showing some of the different room temperature GaAs (100) surface reconstructions that can be observed as a function of surface composition. The relative As/Ga ratio was measured with Auger spectroscopy. A (1x1) surface not shown can also be prepared. (Ref. 26)

Cho showed using HEED that the (111) GaAs face had two phases, a 2x2 reconstruction and a $19^{1/2} \times 19^{1/2}$ R23.4° structure which depended on the relative As to Ga flux incident on the surface. These are termed Ga stabilized and As stabilized.[17] Ranke and Jacobi investigated the compositional dependence of these structures.[24] Auger signals for differently reconstructed (111) surfaces where obtained and related to the composition.

Cho found similar structural results for the GaAs (100) surface.[18] The (100) surface however shows a large number of phases. Arthur[2,20] and later Joyce and Foxon,[25] investigated the compositional dependence of the surface reconstructions using flash desorption measurements. Arthur showed that substantially different As_2 mass desorption curves as a function of temperature were obtained for the As-stabilized and Ga stabilized structures. A peak associated with excess As being driven off was observed and the structure converted to Ga stabilized.

Figure 3 shows LEED photographs obtained by Drathen, et al. which exhibit the range of structures.[26] The phase boundaries of these structures have been studied by Massies, et al. who obtained similar LEED data.[27] The fractional As coverage associated with these phases is shown in Table 1. At the growth temperatures typically used (550-600°C) the dominant phases are the C(8x2) Ga and the C(2x8) As which converts at approximately 1/2 monolayer As excess to deficit. Fig. 4 shows the phase ranges established by Massies, et al.[27]

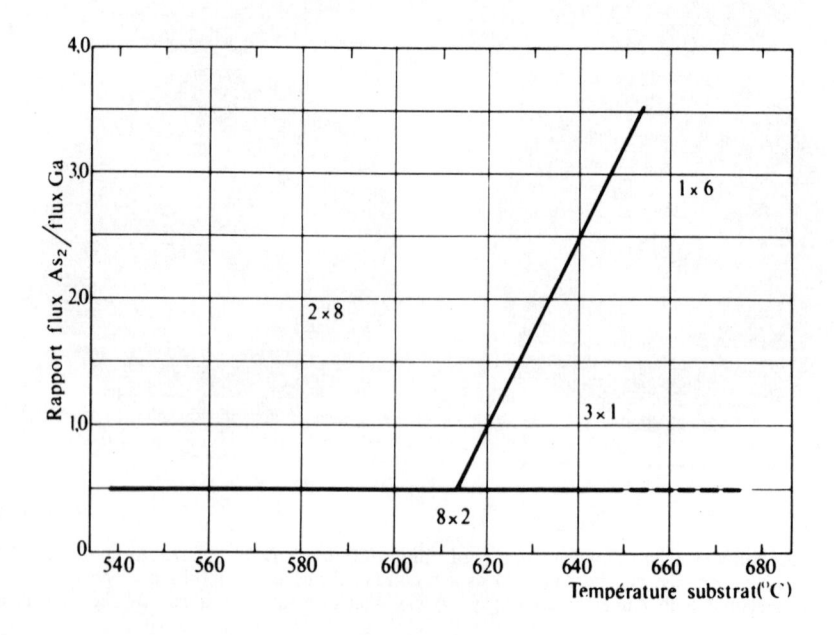

Fig. 4   Phase ranges of surface reconstruction on GaAs (100) in the vicinity of the growth temperature as a function of As_2/Ga flux ratio.   (Ref. 23)

Table I

Comparison of surface structure and composition range for room temperature GaAs (100).

As coverage	Structure Drathen, et al.[26]	Structure Massies, et. al.[23]
0.22	c(8x2)Ga	c(8x2)
		(4x1)
0.27	(4x6)Ga)[1]	(4x6))[a]
0.37	c(6x4)	
0.52	(1x6)As	(1x6)
		(3x1)
0.61	c(2x8)As	c(2x8)
0.86	c(4x4)As	c(4x4)
1.0		(1x1)

)[a] This structure is a superposition of (4x1) and (1x6) domains.

4. Adsorption and Desorption

MBE deposition proceeds via adsorption processes on the crystalline substrate. The primary adsorbates are the elements of which the epitaxial semiconductor are composed, secondarily the evaporated dopants and thirdly the unintentional contaminants. The eventual growth of the crystal on the cleaned substrate proceeds in stages which can include physisorption, chemisorption and chemical reaction. At the temperatures at which semiconductor epitaxy occurs, physisorption binding is weak. If physisorption is the main sticking process, then the surface residence time will be short and no deposition will occur. Chemisorption and chemical reaction are the more important processes. The details of how these proceed depend on the surface structures which have already been introduced. In addition, the nucleation and growth can generate or propogate structural defects.

The primary surface structures discussed in the previous section markedly effect the surface interaction with adsorbates. Three examples of epitaxy can be drawn from GaAs substrates and epitaxy of GaAs,[23] ZnSe[15] and Al[28,29] respectively. The interaction of the elements involved is very different. This depends also on the molecular species in the gas phase. Ga and Al evaporate monatomically, As evaporates typically as As_2 and As_4,[30] Zn as Zn_2[31] and Se as Se_2, Se_4 and Se_8.

The discussion needs to be divided into the primary composition elements and the intentional and unintentional dopants.

4.1 Interaction of As, Ga and Al on GaAs

Growth of GaAs or GaAlAs on GaAs is controlled by the Ga and Al arrival rate at epitaxial growth temperatures since arsenic does not stick. The kinetic processes involved with Ga and As interaction have been reviewed by Joyce and Foxon.[24,32] The relevant details are discussed here. Similar studies although less detailed have been carried out for the II-VI's by Smith.[33]

The initial issue is how the elemental flux interacts with an isolated surface and then secondarily how the interaction is modified in the presence of concurrent fluxes.[1] Fig. 5 shows as an example the desorption of Arsenic from a GaAs (100) surface as a function of surface temperature in the presence of an As_2 flux of 10^{13} mol cm^{-2} sec^{-1}. The source in this case was GaAs. The As_4 then results from an association reaction on the surface.

The desorption energies for As_2 and As_4 have been determined to be 0.38 eV and 0.58 eV respectively. This compares with Ga which has a desorption energy of 2.48 eV for the GaAs (111) face and is strongly chemisorbed. The Ga is mobile, however, and has a surface diffusion range of 200A at the growth temperature.[2] The origin of the chemisorption is seen from photoemission studies of Ga deposited on the (110) GaAs face at room temperature.[34] Fig. 6 shows valence band density of states of GaAs as a function of

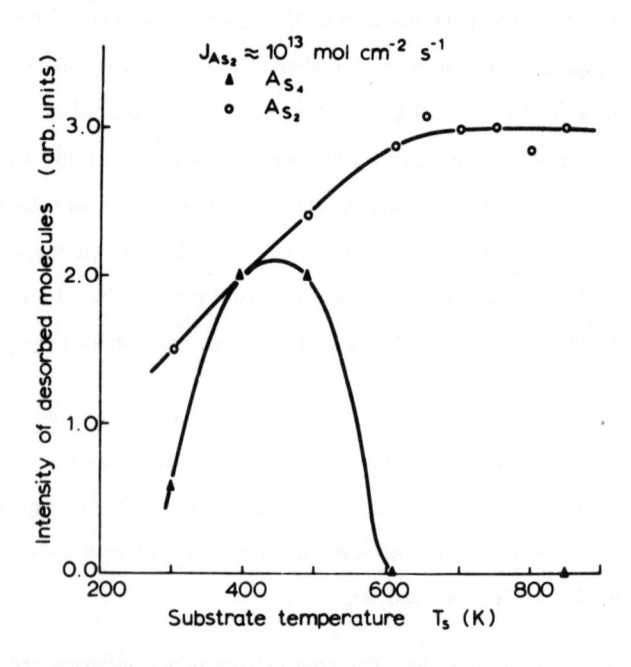

Fig. 5 Relative desorption rates of As_2 and As_4 for an incident As_2 flux as a function of substrate temperature (Ref. 30)

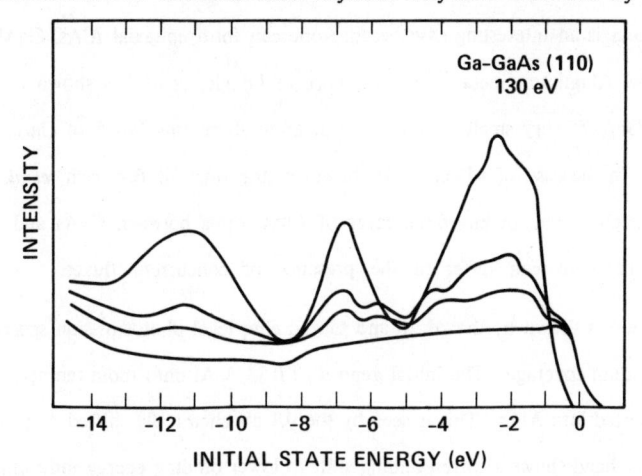

Semiconductor Surface and Crystal Physics Studied by MBE

Fig. 6 Evolution of the valence band photoemission spectra going from cleaved GaAs (110) to a thick Ga metal overlayer. The intermediate exposures represent one-half and one monolayer of Ga on GaAs. The new peaks that arise at -4.2 eV and -5.8 eV reflect bonding across the interface. (Ref. 34)

coverage. In the low coverage range, interface states are observable which indicate that bonding is occurring across the interface. This is accompanied by a charge transfer from the surface As to the Ga and results in an observable As chemical shift shown in Fig. 7.

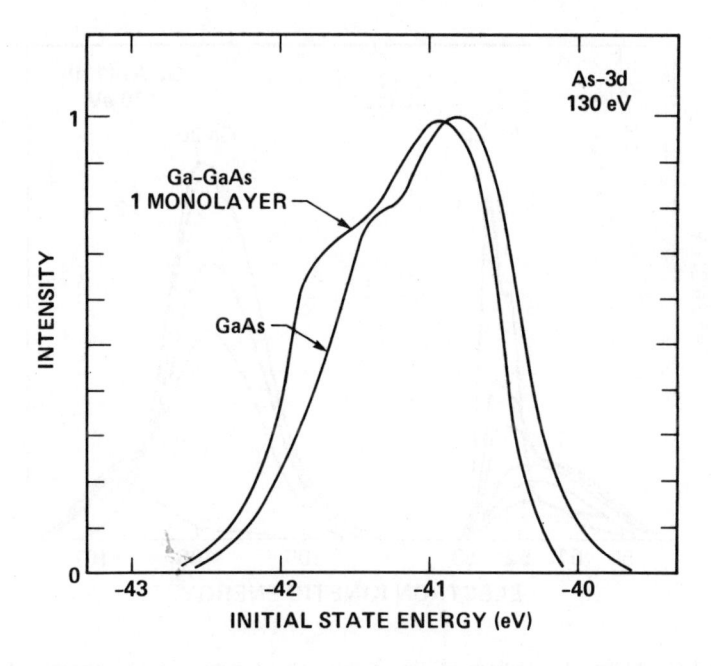

Fig. 7 As 3d chemical shift with Ga coverage on GaAs (110). The shift saturates by one monolayer. (Ref. 34)

Aluminum with GaAs is an interesting case because one can form epitaxial AlAs, GaAlAs ternary alloy or one can epitaxially grow Aluminum metal.[28,29] The work of Dingle, et al. has shown that the interdiffusion coefficient of AlAs in GaAs is very small so that one can grow alternating layers of GaAs and GaAlAs which are stable to 900°C.[35] In the case of Al on GaAs, however, the work of Bachrach, et al. has shown that an exchange reaction takes place and an interfacial layer of AlAs forms between GaAs and Al.[36,37] This is an example of how the reactions can differ in the presence of concurrent fluxes.

The exchange reaction is shown by the Al 2p and Ga 3d core level photoemission spectra presented in Fig. 8 as a function of aluminum coverage. The initial deposit of 0.53 Å Al onto room temperature GaAs (110) was almost completely converted into AlAs. This is seen by the 0.7 eV chemically shifted Al peak to higher binding energy. Ga on the other hand shows a corresponding shift to lower binding energy indicating metallic Ga. The fractional peak intensities are consistent with a complete monolayer of AlAs being formed. The difference between the chemisorption case and the chemical exchange reaction are exemplified in Fig. 9. The heat of reaction which favors this exchange also results in GaAs diffusion into the overlayer. The presence of arsenic during growth would presumably suppress the exchange reaction and allow abrupt heterojunctions to be grown.[35]

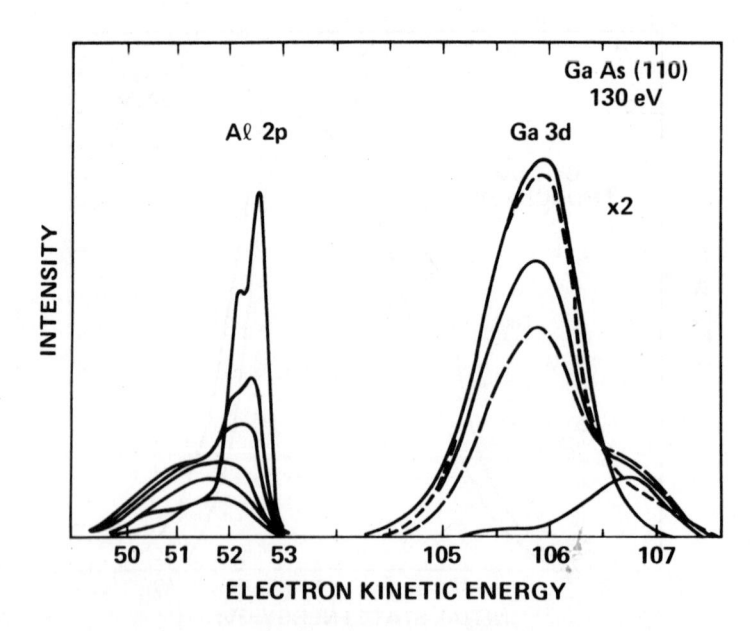

Fig. 8   Al 2p and Ga 3d core level photoemission spectra as a function of Al coverage. The chemical shift indicates an exchange reaction takes place creating an interfacial layer of AlAs and releasing Ga metal into the overlayer. The Al coverage sequence is .53, .93, 1.33, 1.73, 3.73, 16.13 and 88A. (Ref. 37).

Semiconductor Surface and Crystal Physics Studied by MBE

CHEMISORPTION

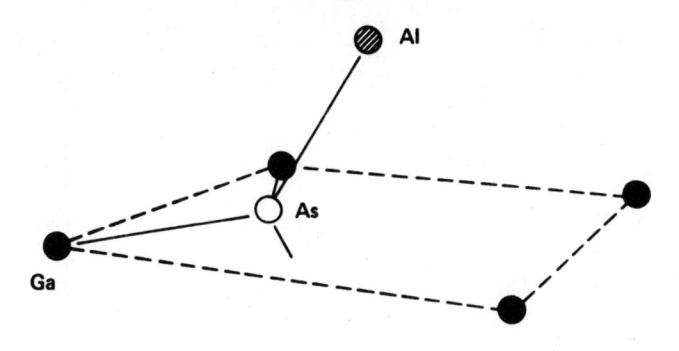

CHEMICAL REACTION

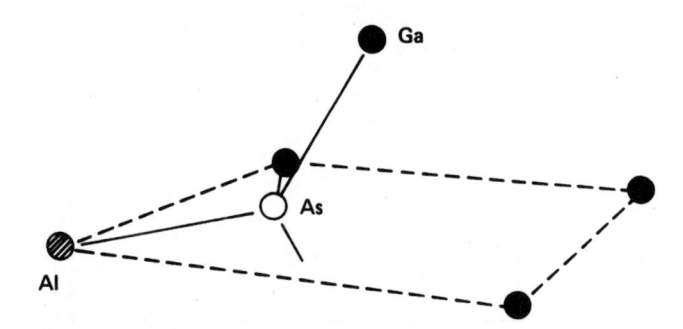

Fig. 9 Schematic of the GaAs(110) unit cell depicting difference for deposition with chemisorption and chemical reaction.

4.2 Doping

Doping effects with MBE for groups III-IV, IV and II-VI semiconductors are complex. Many of the dopants which are useful in other forms of epitaxy do not behave well with MBE. The actual behavior is strongly dependent on substrate temperature and the surface reconstruction involved with growth. The doping regime is typically < 10^{19} cm^{-3}. Above 10^{20} cm^{-3} becomes an alloy regime.

Commonly used dopants in GaAs are Sn, Ge and Si for n-type[38] and Mn and Be for p-type.[39,40] In silicon, Sb is used for p-type and Ga for n type.[41,42] Al does not behave well in Si for profile control because of the strong interdiffusion.[42] The IV-VI's can be doped by control of stoichiometry.[43]

Fig. 10 shows the induced carrier concentration versus reciprocal source temperature, for Be in GaAs, one of the better controlled p type dopants. The actual quality of the crystal obtained is dependent on the surface reconstruction which is exemplified with photoluminescence measurements in Fig. 11. The crystal grown under

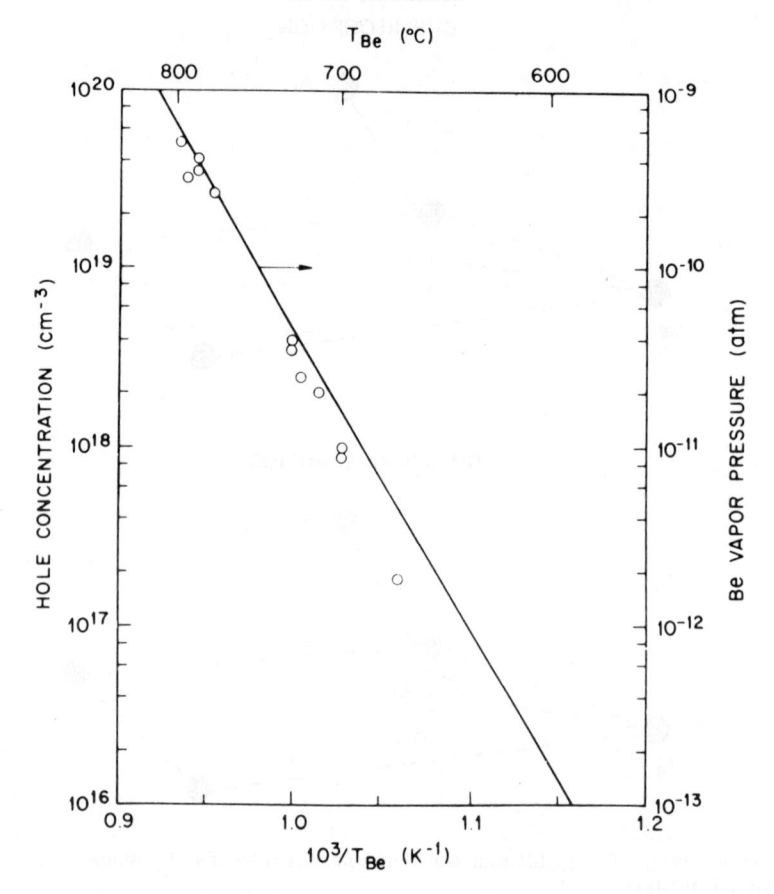

Fig. 10 Hole concentration versus Be effusion cell temperature. The average growth rate was 1.4 μm/hr with a cell to substrate distance of ~5 cm. The solid line represents the Be equilibrium vapor pressure curve. (Ref. 40)

Ga rich conditions is considerably more efficient. This result is consistent with studies of deep traps with capacitative spectroscopy by Lang, et al.[44] Fig. 12 shows an example of such a result, in this case for an n-type sample. In these studies, a variety of unidentified levels appear. The number and concentration, however, are substantially larger when growth occurs under As stabilized conditions. Better understanding of the origin and incorporation of these defects will be an important area of study in the next few years.

The kinetic and interactive aspect of the doping problem is further exemplified by the result of Cho for Mg doping.[45] Fig. 13 shows the increase in "effective" Mg sticking coefficient deduced from the electrically active Mg and corresponding increase in doping level as a function of the Al mole fraction. Joyce and Foxon have shown that in fact Mg has unity sticking coefficient but that on GaAs the Mg diffuses away from the surface and sits at electrically inactive interstitial sites.[32] The co-evaporation of Al modifies the tendency for Mg to sit interstitially, but the mechanism has not been established.

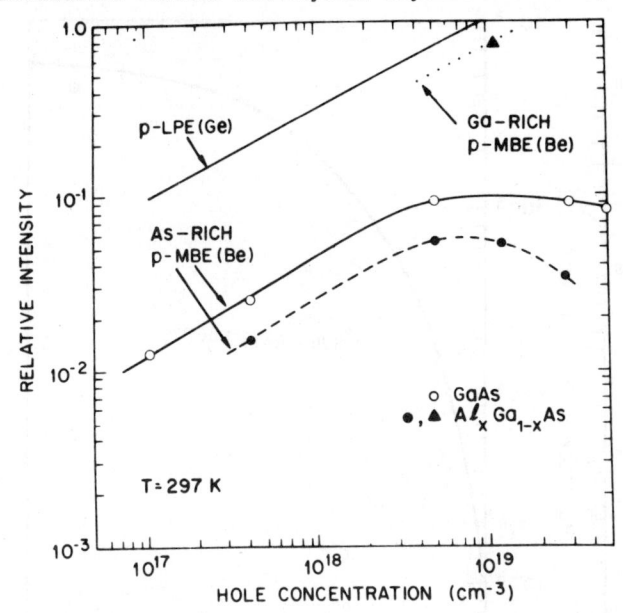

Fig. 11 Relative photoluminescence intensity for Be doped GaAs (O) and $Al_{0.17}Ga_{0.7}As$ (•) layers grown under As-rich conditions and for $Al_{0.17}Ga_{0.83}As$ (△) layer grown under Ga-rich conditions. The results for Ge-doped LPE GaAs are shown as a reference. (Ref. 40).

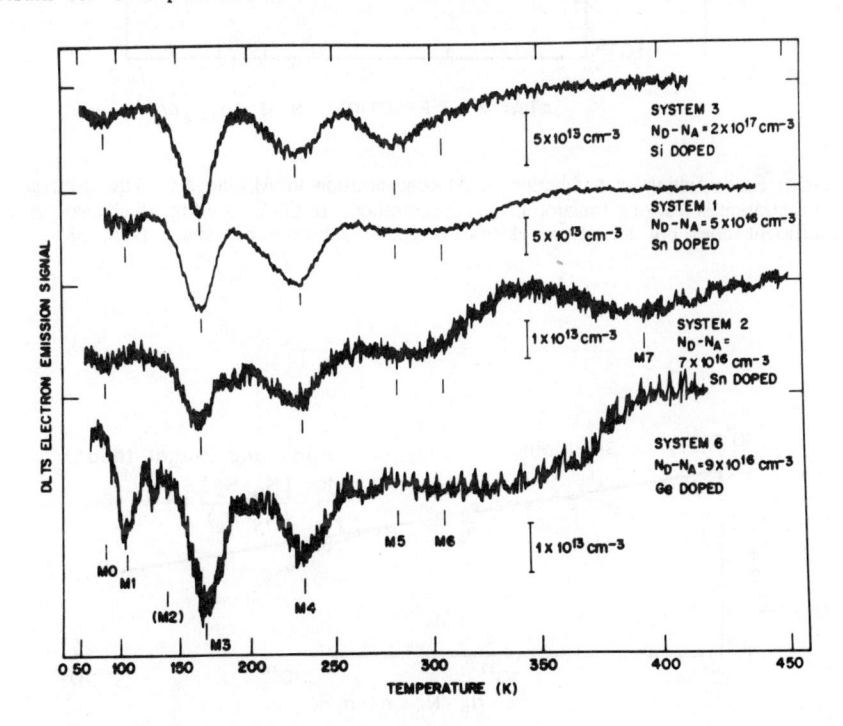

Fig. 12 DLTS capacitance spectra of electron traps in n-GaAs. The two lower curves are for samples grown under As-rich conditions and the upper curve was for a sample grown under Ga-rich conditions. Each of the labeled peaks represents a different impurity or defect related center (Ref. 44).

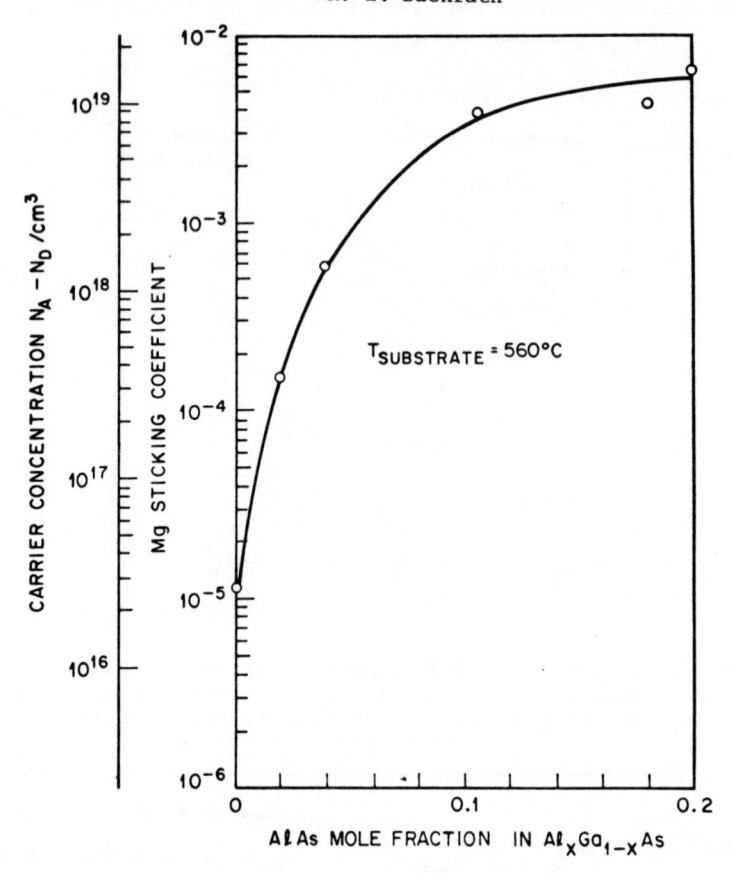

Fig. 13 Carrier concentration as a function of Al concentration in $Al_xGa_{1-x}As$. The electrically active Mg concentration is a strong function of Al concentration. In GaAs, the Mg which also has unity sticking coefficient does not sit at an electrically active substitutional site. (Ref. 38)

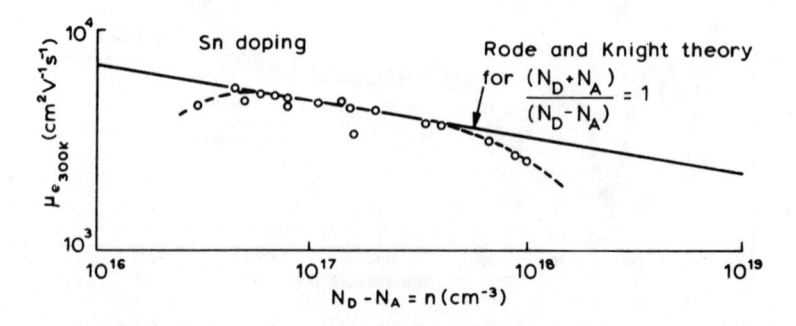

Fig. 14 Electron mobility as a function of net donor concentration (N_D-N_A) for Sn doped films. Deviation from the theoretical line is an indication of the level of compensation. (Ref. 32)

An opposite form of behavior occurs for Sn which has a tendency to accumulate at the surface.[38,46] Sn doping has been well studied and results in a uniform doping level.[47] In the range 5×10^{16} - 5×10^{17} carriers cm^{-3} the compensation level was low.[32] This was established by comparing the mobility as a function of carrier concentration with the calculation of Rode and Knight[48] as is shown in Fig. 14.

The high concentration limit of the doping results in alloys or overlayers. These are discussed in more detail in connection with the discussion of surface and interface states.

Two other examples of doping are the interaction of zinc[31] with GaAs and the group VI elements which in the bulk form neutral donors. O, S, Se and Te when evaporated typically evolve as molecular species which at the growth temperature interact weakly with the surface. This low sticking coefficient leads to low doping. These effects are strongly temperature dependent. Studies of the oxidation of GaAs are protypical of these doping phenomena and are discussed in the next section.

Zinc is one of the primary p-type dopants in LPE grown GaAs, but for MBE, the sticking coefficient is so low that no appreciable doping can be achieved with a thermal evaporation source. By using an ionized source, electrically active zinc can be incorporated.[49] This is shown in Fig. 15 where a sticking coefficient of 0.03 is deduced. The incorporation has been explained as an ion implanation result. The actual mechanism, however, has not been clarified since thermally evaporated zinc is presumably composed of Zn_2 while the ionization would probably break up these dimers. A similar approach has been used for nitrogen doping in GaAsP.[50]

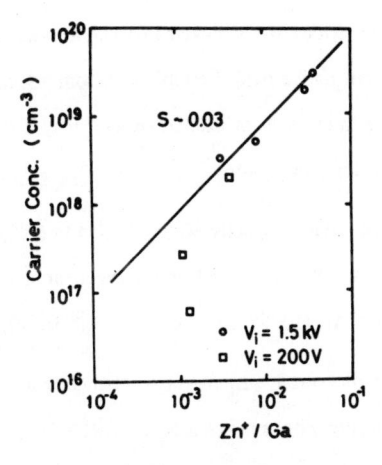

Fig. 15 Hole carrier concentration as a function Zn^+/Ga arrival rates. The Zinc ion accelerating voltages were 1.5 KV (O) and 200 V (□). The solid line corresponds to an effective sticking coefficient of 0.03. Without the ionizer, the sticking coefficient is ~0. (Ref. 49)

4.3 Oxidation

A discussion of oxidation is relevant because the interactions are typical of the doping reactions for other Group VI elements in GaAs grain by MBE. The group VI elements are not useful for dopants in GaAs. Sulphur, selenium and tellurium have a weak interaction with GaAs and result in few incorporated impurities at the growth temperature. This results from the competing desorption reaction.

Oxygen in GaAs forms a deep substitutional donor with low solubility and on the surface at high concentration can form a native oxide. Studies of oxidation of semiconductors prepared *in-situ* with MBE provide access to other than the cleavage faces. Examples of this type of work are studies of the GaAs (100), (111)[21] and (110)[51] faces, and the ZnSe (100) face.[15] A related problem is the oxidation of Ga[36] and Al metal.[52,53] For example, a monolayer coverage of oxygen on GaAs requires a 10^6 Langmuir exposure while ~10^2 L on Al and 10^3L on Ga provides a saturated coverage. In the MBE growth environment, the oxidation of the metals prior to reaction into the semiconductors is the more likely step if oxygen or oxygen containing molecules are present. When an oxygen beam is present during the growth of GaAlAs, semi-insulating material can be grown.[54]

With reference to Fig. 2, one sees that the unreconstructed polar (100) and (111) face have surfaces with a specific outermost atom. The (110) surface as discussed earlier undergoes a relaxation whereby the Arsenic moves out and the Ga moves in so that the three faces are structurally not all that different, so the compositional effects might be more important.

The initial oxidation of GaAs has been studied in detail both on MBE grown GaAs[21,24,51] and on *in-situ* cleaved GaAs (110).[55] This oxide depends on whether the exposure is to molecular oxygen or to excited oxygen. Compositionally only very thin oxides are achieved in vacuum without plasma activation. These thin oxides seem to differ compositionally from native oxides established from aqueous solution,[56] but this discussion is beyond the scope of this article. The interaction of excited oxygen may also relate to the higher sticking coefficient of zinc evaporated with an ionizing source.

Figure 16 and 17 show oxygen uptake data respectively for the GaAs (100)[21] and (111)[24] face as a function of oxygen dose. In both cases, the Ga rich surface adsorbs oxygen at a much faster rate. These can be compared with the uptake curves for Al (100)[52] and Ga[36] metal in Figs. 18 and 19.

Fig. 16 shows oxygen uptake curves measured with Auger spectroscopy as the ratio of the oxygen to gallium signals. These exposures are made statically with a maximum pressure of 10^{-4} Torr (note that 10^{-4} Torr min = 6×10^3 Langmuirs). Initially a gallium rich surface chemisorbs oxygen at a much faster rate.

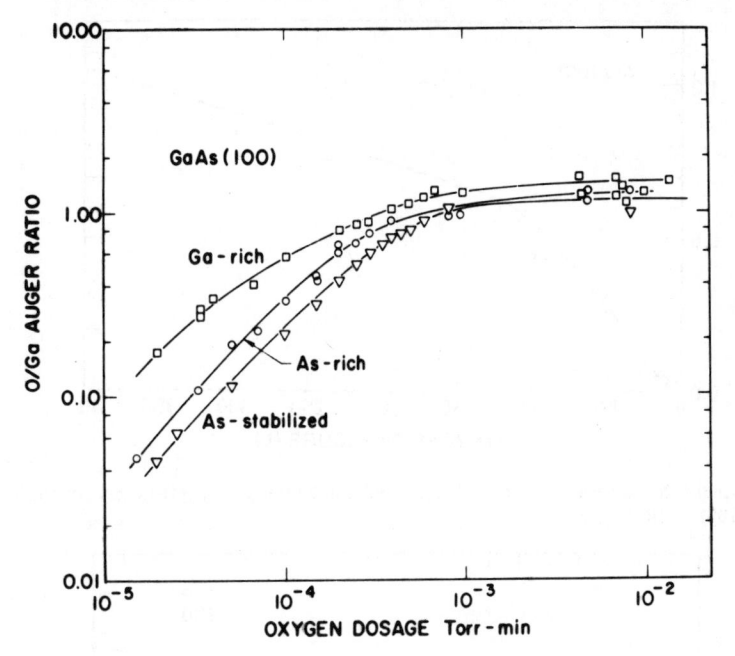

Fig. 16 Oxygen uptake measured with Auger spectroscopy room temperature oxygen exposure curves are for Ga-rich (4x6), As-stabilized c(2x8), and As-rich (1x1) surface structures. (10^{-3} Torr-min = 6×10^4 L.) (Ref. 21)

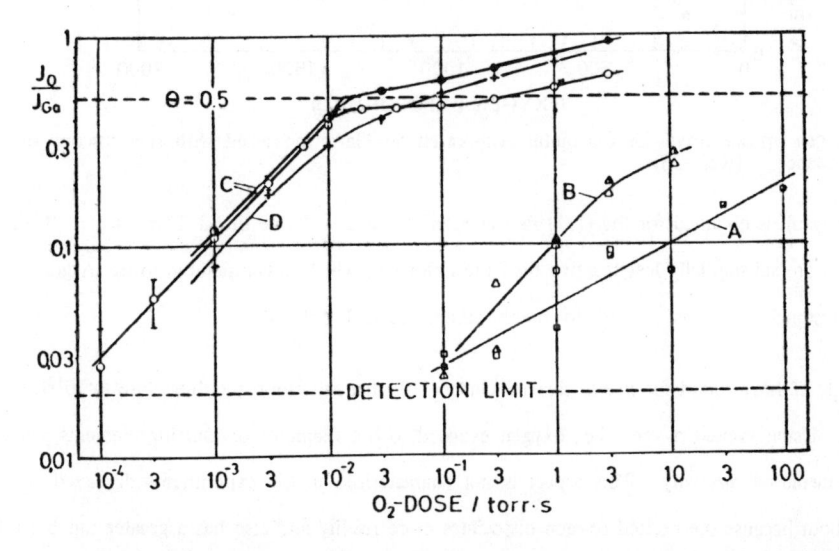

Fig. 17 Oxygen uptake curves measured with Auger spectroscopy for GaAs (111) surfaces. (A) As-stabilized (2x2); (B) Ga-stabilized ($19^{1/2} \times 19^{1/2}$); (C) Argon bombarded and annealed for 10 min. at 500°C (two separate runs); (D) Argon bombarded only. (1 Torr-sec. = 10^6 L.) (Ref. 24)

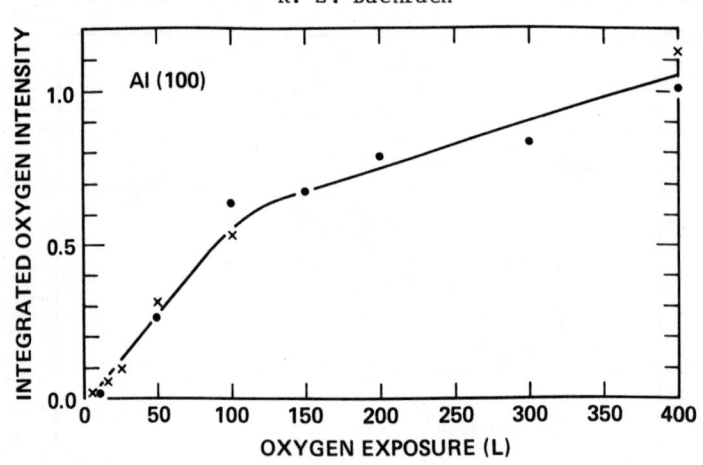

Fig. 18 Oxygen uptake curve measure with valence band photoemission spectroscopy for the O_2 p resonance on Al (100). (Ref. 52).

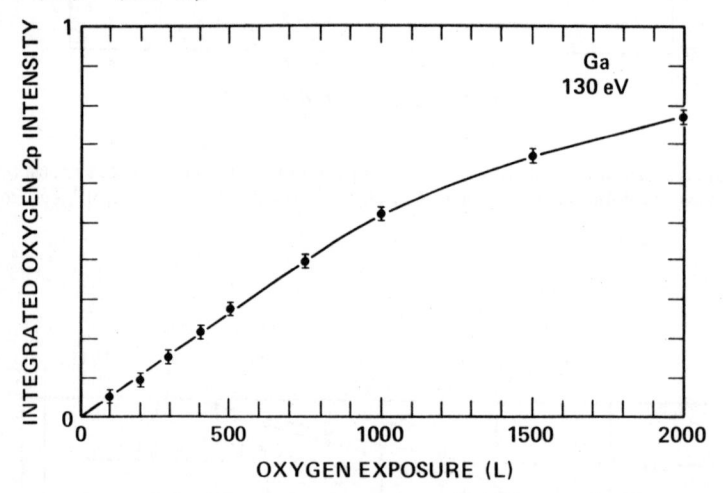

Fig. 19 Oxygen uptake curve for Ga metal evaporated on GaAs measured with spectroscopy of the O 2p resonance. (Ref. 36)

A similar result is obtained for the (111) face as seen in Fig. 16. (Note that 1 Torr - sec = 10^6L.) The as grown surfaces are substantially less reactive than the (100) face which is comparable to the Argon bombarded surfaces. Presumably this is related to bombardment related defects.

Pianetta, et al. have found for in-situ cleaved GaAs (110) that saturation requires close to 10^{12}L of ground-state oxygen. Using excited oxygen, i.e., oxygen exposed to hot filaments or ionizing elements, can result in substantially increased reactivity. This aspect is not characterized in the experiments discussed above. The effect arises both because the excited oxygen dissociates more readily and also has a greater capability to break surface back bonds. Pianetta also found for the (110) surface with UPS that the O chemisorbed to arsenic. This result led to some controversy since other ELS evidence suggested that O chemisorbed to gallium.

The evidence for gallium involvement on the (111) and (100) faces is particularly strong, but in these cases there is some evidence that molecular oxygen chemisorbs. Evidence for the (110) face suggests in this case the chemisorption is dissociative. Desorption measurements typically show a gallium oxide evolution,[57] however, the question of surface mobility needs to be investigated more. If the oxygen becomes mobile, it has a greater ability to break a Ga bond than an As bond. Consistent with this, the sticking coefficient also decreases with increasing As content. Charge transfer certainly occurs from the surface As, but the structure and occurrence of special sites is more important for the adsorption of oxygen then the surface concentration of arsenic. The calculation of Mele and Joannopoulos[58] has clarified many of the issues relating to the interpretation of UPS and ELS studies. Their result for the (110) surface indicates that oxygen adsorption can produce spectral changes in both arsenic and oxygen derived features. Their calculation indicates a preferential chemisorbed bond to the surface arsenic. More theoretical work is called for with respect to the polar faces.

As discussed earlier, the oxide formed in GaAs desorbs at high temperatures.[57] Among the studies investigating oxide sublimation, Ranke has used photoemission at 21.2 eV.[59] Fig. 20 shows spectra taken for the GaAs (111) face. The upper curve is the clean surface and the lower difference curves show the behavior as a function of temperature of a saturation exposed surface. By 870 K, the clean surface is substantially restored. The major peak at -5 eV is an O 2p related surface resonance which is characteristically seen when oxygen adsorbs to a surface. This transition has a strong cross section and is generally useful for studying oxidation related effects.

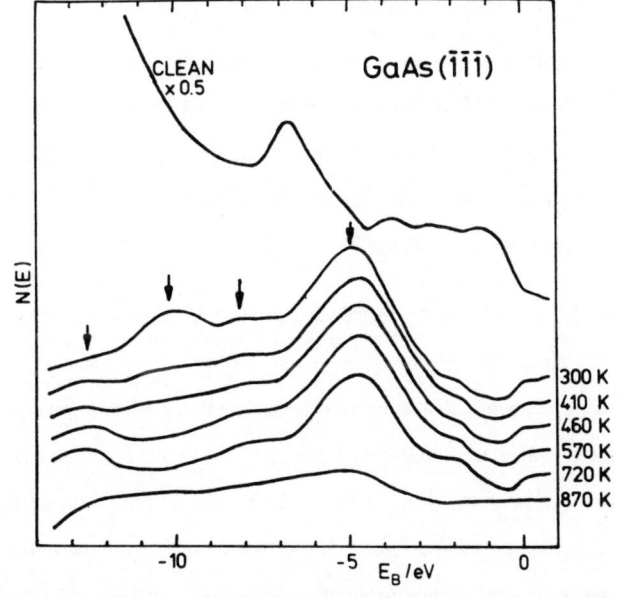

Fig. 20 GaAs (111) photoemission valence band spectra measured with 21.2 eV excitation. The upper curve is the clean semiconductor while the lower curves are different spectra measured at room temperature after raising the oxygen exposed surface to the temperature shown. (Ref. 59)

Studies of the oxidation of GaAlAs are only beginning, but preliminary results indicate that the Al strongly chemisorbs O.[37] This is to be expected based upon studies of the oxidation of aluminum metal. The question of the oxidation of GaAlAs is important to the growth of heterostructure laser devices and the quality of the GaAlAs-GaAs interface.

MBE has also been used to investigate the oxidation of ZnSe(100)[15] epitaxially grown on GaAs. GaAs and ZnSe lattice match well, but during growth the ZnSe surface exhibits a (2x2) reconstruction. When exposed to oxygen, the ZnSe does not show a gradual uptake similar to GaAs. Rather, above a critical pressure of 0.08 Torr, an abrupt exchange reaction takes place. This seems to be accompanied by a loss of selenium from the surface.

5. Surface and Interface States on Semiconductors

Intrinsic surface states on semiconductors as well as their modification with both metal-semiconductor and semiconductor-semiconductor junction formation have been of renewed interest. Molecular beam epitaxy has provided better control to these studies by allowing access to a variety of atomically clean surfaces. The evaporation control also allows the interfaces that form to be investigated at various stages. In this section representative examples of this type of study will be presented. The spectroscopic techniques which have been employed for these types of studies are photoemission and electron energy loss. These probe respectively the occupied and unoccupied density of states.

Extensive angle integrated and angle resolved photoemission measurements have been carried out on cleaved GaAs (110) surfaces.[60-64] Similar measurements have not yet been made for the faces typically grown *in-situ* by MBE and as a function of surface reconstruction, although work of this type is beginning. Ludeke has reported angle integrated MgK excited photoemission studies of the GaAs (100) face taken at grazing incidence to enhance the surface sensitivity.[65] These have shown new apsects of the surface states.

Different aspects of the surface and interface states relate to various important phenomena. Those states which arise below the top of the valence band in semiconductors are involved with bonding. The surface localized charge or dangling bonds that these represent are often the states with which an adsorbate is most likely to interact. States within the fundamental gap can directly effect conductivity, transport and electronic barriers that may form. For many years it was thought that intrinsic surface states in the gap played a fundamental role in Schottky barrier formation. The extrinsic aspects of these effects is better understood now and the manner in which they are related to interface formation, reactions and the charge transfer that occurs.

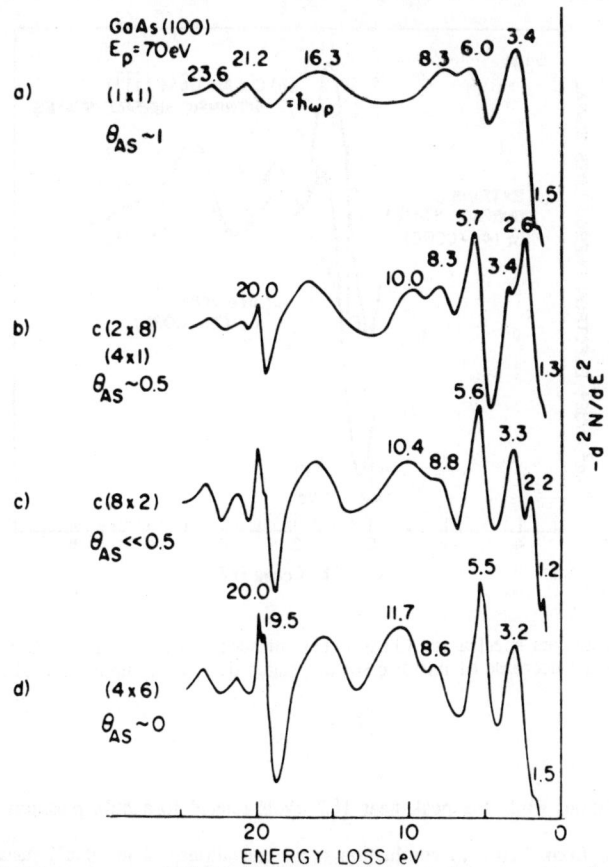

Fig. 21 Electron energy loss spectra of GaAs (100) for different surface reconstructions. $E_p = 70$ eV is the primary electron energy. (Ref. 21)

Unoccupied states below the conduction band edge can act as recombination centers and if the density is large, pin the Fermi level upon Schottky barrier formation. States above the conduction band edge can mediate charge transfer from adsorbates.

Unoccupied states can be probed with several techniques. One commonly used is electron energy loss usually plotted as second derivative spectra. This spectroscopy is analogous to Raman spectroscopy although the selection rules are different. Features arise at energies shifted from the primary energy by the interstate transition energy. The photon excited technique of constant final state partial yield spectroscopy also probes these transitions.[64] They gain their surface sensitivity because they are probing with electrons with 1-100 eV kinetic energy.

Fig. 21 shows energy loss spectra for GaAs (100) as a function of As coverage and thereby the corresponding reconstruction.[20] Features below 20 eV have valence band related initial states while above 20 eV, transitions also occur from the Ga 3d core level. The peaks near 3.4 and 5.6 eV are attributed to bulk

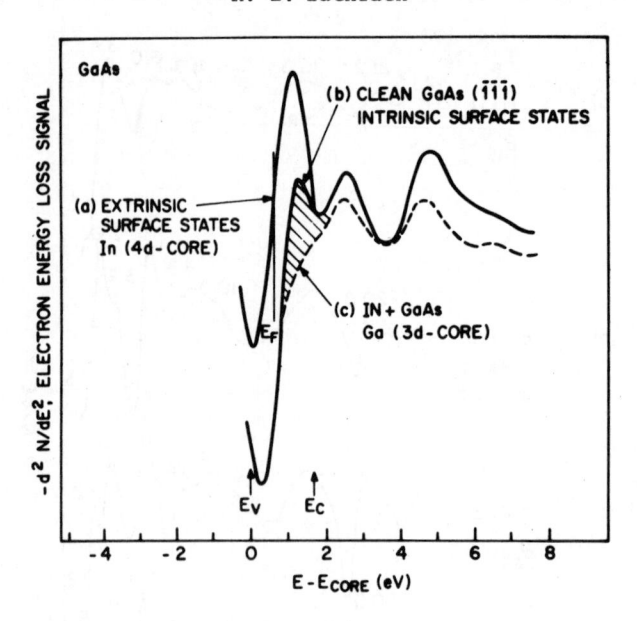

Fig. 22 Electron energy loss spectra for GaAs (111) surfaces in the vicinity of the Ga 3d core level. The spectra show the decrease of intrinsic surface states due to ~1 monolayer of indium atoms. (Ref. 65)

related interband excitations while the peak near 16.3 eV is related to a bulk plasmon creation. The peaks at 21.2 and 23.6 eV arise from d-core to conduction band transitions. The 20 eV peak on the other hand is associated with a d-core to empty dangling bond states principally associated with the Ga atom. The strong peak at 2.6 eV which occurs for intermediate coverages is due to excitations from filled As dangling-bond states to empty Ga dangling bond states.

These transitions are modified by the presence of overlayers. Rowe, et al.[66] have investigated In on GaAs (111) as shown in Fig. 22 and Ludeke, et al. have investigated both oxidation with this technique and Sb overlayers on GaSb (100), GaAs (100) and InAs (100).[67] These results are shown in Fig. 23. Similar overlayer studies have been performed by Massies, et al. for Ag on GaAs (100) and InP(100).[27]

The data in Fig. 21 indicate empty surface states localized about the Ga atom. For small coverages of metal atoms or ~1/2 to 1 monolayer, these empty states are removed and the only empty states observable by ELS occur near the metal atom. The ELS line shape for these thin metal layers is considerably sharper than that for thick ~1200 Å metal films and this suggests a more covalent type of bonding for the interface metal atoms than for the bulk metal. A similar conclusion has been reached for aluminum surfaces and this shows up in the charge transfers that occur for Al on GaAs.[52]

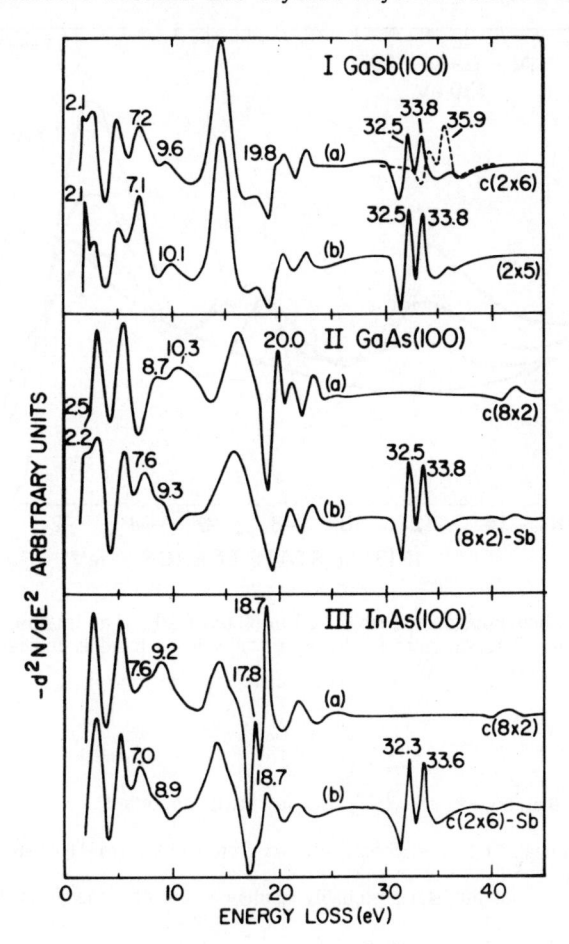

Fig. 23 Electron energy loss spectra for clean and Sb covered GaSb, GaAs and InAs (100) surfaces. The spectra show the development of interface states with the overlayer. (Ref. 66).

Another aspect of interface states is seen in Fig. 24. The spectra show the as grown surfaces and then Sb covered.[66] The important features are the Sb derived transition to empty surface states which are located close to the cation-derived surface states near the bottom of the conduction-band edge. These states are specific to the (100) surfaces and arise because of a rehybridization of the two sp^3 dangling orbitals.

Substantial work investigating the occupied density of states is not yet available, although studies are beginning. Fig. 26, for example, shows valence band changes that occur with Al overlayers on GaAs.[37] These spectra show the transition from clean GaAs (100) to Al metal. The structure in the Al metal valence band would indicate epitaxy was achieved.[52]

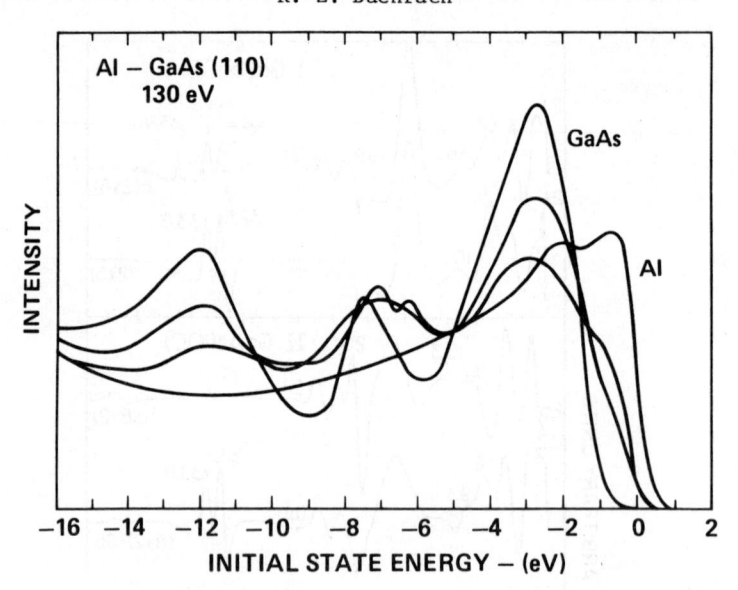

Fig. 24 Valence band photoemission spectra for Al on GaAs (110). The intermediate coverages are one and two monolayers. The structure in the Al metal valence band is indicative of epitaxial deposition. (Ref. 37)

The interface state that occurs at -6.1 eV is particularly important for it reflects bonding between the semiconductor and the metal. This is different however from the Ga overlayer spectra shown in Fig. 6. The absence of the -4.2 eV back bonding states probably results because of a change in the surface relaxation which pushes these states to lower binding energy.

Further investigation of the inteplay of surface structures and surface electronic states will be an active topic in the next few years. The experimental possibilities created by MBE should allow new and more detailed experiments to be performed.

References

1. K. G. Gunther, "Interfacial and Condensation Processes Occurring with Multicomponent Vapors", p. 213, "The Use of Thin Films in Physical Investigations" edited by J. C. Andersen, Academic Press (1966).

2. A. Y. Cho and J. R. Arthur, Progress in Solid State Chemistry, 10, 157 (1975). This work contains extensive references and a bibliographic review.

3. L. L. Chang and R. Ludeke, p. 37, Epitaxial Growth, Part A, edited by J. W. Matthews, Academic Press (1975); and L. Esaki and L. L. Chang, Chemistry and Physics of Solid Surfaces, Edited by R. Vanselow and S. Y. Tong, CRC Press, p. 111 (1977); RFC Farrow, 1976 Crystal Growth and Materials, Edited by E. Kaldes and H. J. Scheel, North Holland Pub. Co., p. 238 (1977).

4. P. E. Luscher and D. G. Collins, J. Crystal Growth and Characterization, this issue.

5. R. Z. Bachrach, L. E. Swartz, R. D. Burnham and A. Alimonda, J. Vac. Sci. and Techn., to be published.

6. J. R. Arthur, CRC Critical Reviews in Solid State Sciences, p. 416, August (1976).

7. J. P. Hobson, Japan J. Appl. Phys. Suppl. 2, p. 317 (1974).

8. D. W. Pashley, "A Historical Review of Epitaxy", p. 1; Epitaxial Growth, Part A, edited by J. W. Matthews, Academic Press (1975).

9. D. B. Dove, "High-energy Electron Diffraction", p. 331 - Epitaxial Growth, Part A, Edited by J. W. Matthews, Academic Press (1975).

10. P. W. Palmberg, Anal. Chem., 45, 549A (1973); and L. L. Chang, Surf. Sci. 25, 53 (1971).

11. K. Ploog and A. Fischer, Appl. Phys. 13, 111 (1977); C. E. C. Wood and B. A. Joyce, J. Appl. Phys. 49, 4854 (1978).

12. J. C. Bean, G. E. Becker, P. M. Petroff, and T. E. Seidel, J. Appl. Physics, 48, 907 (1977).

13. C. B. Duke, Crit. Rev. Solid State and Material Sciences, 8, (1978); and J. A. Appelbaum and d. R. Hamann, Chemistry and Physics of Solid Surfaces, Edited by R. Vanselow and S. Y. Tang, CRC Press, p. 275 (1977).

14. J. B. Pendry, Low Energy Electron Diffraction, Academic Press (1974).

15. R. Ludeke, Solid State Comm., 24, 725 (1977).

16. J. E. Rowe and H. Ibach, Phys. Rev. Lett., 32, 421 (1974).

17. D. J. Chadi, Phys. Rev. Letters, to be published.

18. A. Y. Cho, J. Appl. Phys. 42, 2074 (1971).

19. A. Y. Cho, J. Appl. Phys. 41, 2730 (1970).

20. J. R. Arthur, Surface Science, 43, 449 (1974).

21. R. Ludeke and A. Koma, CRC Critical Rev. Solid State Sci., 5, 259 (1975).

22. K. Jacobi, G. Steinert and W. Ranke, Surface Science, 57, 571 (1976).

23. J. Massies, P. Etienne and N. T. Linh, Thomson-CSF Technical Review, 8, March (1976).

24. W. Ranke and K. Jacobi, Surface Science, 63, 33 (1977).

25. B. A. Joyce and C. T. Foxon, Journal of Crystal Growth, 31, 122 (1975).

26. P. Drathen, W. Ranke and K. Jacobi, Surface Sci., 77, L162, (1978).

27. J. Massies, P. Devoldere and N. T. Linh, J. Vac. Sci. Technol., 15, Aug/Sept. (1978).

28. R. Ludeke, L. L. Chang and L. Esaki, Appl. Phys. Lett., 23, 201 (1973).

29. A. Y. Cho and P. D. Dernier, J. Appl. Phys. 49, 3328 (1978).

30. C. T. Foxon and B. A. Joyce, Surface Science, 64, 293 (1977).

31. J. R. Arthur, Surface Science, 38, 394 (1973).

32. B. A. Joyce and C. T. Foxon, Jap. Journal of Appl. Physics, 16, 17 (1976).

33. D. L. Smith and V. Y. Pritchard, J. Appl. Phys., 46 2366 (1975).

34. R. Z. Bachrach and A. Bianconi, J. Vac. Sci. Tech., 15, 525 (1978).

35. R. Dingle, Advances in Solid State Physics, Edited by H. J. Queisser, Pergamon Press, p. 21 (1975).

36. R. Z. Bachrach, J. Vac. Sci. Tech., 15, Aug/Sept. (1978).

37. R. Z. Bachrach, R. S. Bauer, J. C. McMenamin and A. Bianconi, Proc. 14th Int. Conf. on Phys. of Semiconductors, Edinburgh, Scotland (1978).

38. A. Y. Cho, J. Appl. Phys., 46 1733 (1975).

39. M. Illegems, R. Dingle and L. W. Rupp, Jr., J. Appl. Phys., 46, 3059 (1975).

40. M. Illegems, J. Appl. Phys. 48, 1278 (1977).

41. G. E. Becker and J. C. Bean, J. Appl. Phys., 48, 3395 (1977).

42. Y. Ota, J. Electrochem. Soc. 124, 1795 (1977).

43. D. L. Smith, J. Crystal Growth and Characterization, this issue.

44. D. V. Lang, A. Y. Cho, A. C. Gossard, M. Illegems and W. Wiegmann, J. Appl. Phys., 47, 2558 (1976).

45. A. Y. Cho and M. B. Panish, J. Appl. Phys. 43, 5118 (1972).

46. K. Ploog and A. Fischer, J. Vac. Sci. Tech., 15, 255 (1978).

47. G. Abstreiter, E. Bauser, A. Fischer and K. Ploog, Applied Physics, (1978).

48. D. L. Rode and S. Knight, Phys. Rev. B3, 2534 (1971).

49. N. Matsunaya, M. Naganuma and K. Takahashi, Jap. Journal of Appl. Phys., 16, 443 (1976).

50. Y. Matsusima, S. Gonda, Y. Makita, and S. Mukai, J. of Cryst. Growth, 43, 281 (1978).

51. R. Ludeke, Solid State Comm., 21, 815 (1977).

52. R. Z. Bachrach, S. A. Flodstrom, R. S. Bauer, S. B. M. Hagstrom and D. J. Chadi, J. Vac. Sci. Tech. 15, 488, (1978);

53. S. A. Flodstrom, R. Z. Bachrach, R. S. Bauer, and S. B. M. Hagstrom, Phys. Rev. Lett., (1978).

54. H. C. Casey, Jr., A. Y. Cho and E. H. Nicollian, Appl. Phys. Lett., 32, 678, (1978).

55. P. Pianetta, I. Lindau, M. Garner and W. E. Spicer, Phys. Rev. B, to be published.

56. C. C. Chang, R. P. H. Chang and S. P. Murarka, J. Electrochem. Soc., 128, 481 (1978); and B. Schwartz, F. Ermanis and M. Bradstad, J. Electrochem. Soc., 123, 1089 (1976).

57. J. R. Arthur, J. Appl. Phys. 38, 4024 (1967).

58. E. J. Mele and J. D. Joannopoulos, Phys. Rev. B (1978), to be published; E. J. Mele and J. D. Joannopoulos, Phys. Rev. Lett., 40, 341 (1978).

59. W. Ranke, Proc. 7th Intern. Vac. Congr. and 3rd Intern. Conf. on Solid Surfaces, 1145, Vienna (1977).

60. W. Gudot and D. E. Eastman, J. Vac. Sci. Technol., 13, 831 (1976).

61. J. A. Knapp and G. J. Lapeyre, J. Vac. Sci. Technol., 13, 757 (1976).

62. P. E. Gregory and W. E. Spicer, Phys. Rev., B12, 2370 (1975).

63. R. Z. Bachrach, R. S. Bauer and S. A. Flodstrom, Photoemission From Surfaces, Edited by R. F. Willis, B. Feuerbacher, B. Fitton and C., Backx, ESTEC, Noordwijk, The Netherlands, 103 (1976).

64. R. S. Bauer, R. Z. Bachrach, S. A. Flodstrom and J. C. McMenamin, J. Vac. Sci. Technol., 14, 378 (1977).

65. R. Ludeke, Proc XIVth Int. Conf. on the Physics of Semiconductors, Sept. 1978, Edinburgh, Scotland.

66. J. E. Rowe, S. B. Christman and G. Margantondo, Phys. Rev. Lett. 35, 1471 (1975); J. E. Rowe, J. Vac. Sci. Technol., 13, 798 (1975).

67. R. Ludeke, Phys. Rev. Lett., 39, 1042 (1977).

THE AUTHOR

R. Z. Bachrach

Robert Bachrach received his Bachelors of Science in Physics from M.I.T. and his Ph.D. from the University of Illinois, Urbana. He was a member of the Technical Staff at Bell Laboratories, Murray Hill from 1969 to 1973 whre his principal studies involved materials characterization and optical studies on GaP and GaAs. In 1973 he joined the Xerox Palo Alto Reserch Center where he is now a Senior Member of the Research Staff. In 1978, Dr. Bachrach received an appointment as a consulting Professor at Stanford University in the Synchrotron Radiation Laboratory. His research has included fundamental studies of soft x-ray physics and the use of synchrotron radiation for studies of metal and semicondcutor surfaces. Molecular Beam Epitaxy studies are being pursued both from the materials and device aspects and as a tool for studying polar surfaces on semiconductors. Dr. Bachrach is a fellow of the American Physical Society and Sigma Xi and a member of the American Vacuum Society, the Optical Society of America and the American Academy for the Advancement of Science. His wife Virginia is a pediatrician and they have three daughters.

PERIODIC DOPING STRUCTURE IN GaAs

G. H. Döhler and K. Ploog

Max-Planck-Institut für Festkörperforschung, Heisenbergstrasse 1,
D-7000 Stuttgart 80, W. Germany

(Submitted January 1979)

ABSTRACT

A new type of superlattice consisting of a periodic sequence of ultrathin p- and n-doped semiconductor layers, possibly with intrinsic (i-) layers in between, is considered. In addition to the features related to the subband formation, as known from the study of the familiar heterostructure superlattices of the type $Ga_xAl_{1-x}As$, crystals with doping superstructure also show other intriguing pecularities. A semiconductor with doping superstructure, in fact, represents a crystal whose band gap and carrier concentration can be tuned within wide limits. In the first part of this review, some of the unique features of this class of materials are discussed. In the second part, approaches for the realization of doping structures and nipi superlattices in GaAs by the technique of molecular beam epitaxy are outlined.

1. INTRODUCTION

The fascinating prospects which result from the possibility to fabricate man-made superlattices with a compositional superstructure have been exhibited in the preceeding article. In this chapter a quite different approach for generating a periodic superstructure potential in a semiconductor will be discussed. Instead of a periodic variation of alloy composition a sequence of alternating n- and p-type ultrathin doping layers, possibly with intrinsic (i-)layers in between, will be considered in the following.

The periodic potential in these doping superlattices (also called "nipi crystals"[1,2], according to the sequence of doping layers) results from the space charge of a relatively small concentration of ionized impurity atoms on an otherwise unperturbed homogeneous bulk. This space charge potential induces a periodic and parallel modulation of the conduction and valence band edges. The key feature of such a system is the local separation between the low lying electron states in the conduction band and the uppermost hole states in the valence bands which are confined to the n- and p-doped layers, respectively. The resulting low, or, depending on the choice of the parameters, even vanishing recombination rate between electrons and holes leads to a large variety of novel features of nipi crystals. Such crystals, in fact represent semiconductors whose effective band gap and carrier concentration may be tuned arbitrarily within widest limits.

After a brief description of the system in section 2.1 various consequences which result from the just mentioned features of a nipi crystal and from its highly anisotropic and dynamically 2-dimensional character will be outlined in section 2.2. The molecular beam epitaxy represents the ideal tool also for the realization of doping superlattices. In particular, the relatively low temperatures required for high quality crystals restricts interdiffusion of the dopants and the high flexibility of the MBE technique makes feasible complicated doping profiles including more complex structures with modulation of the doping in three dimensions.

The section 3 is dealing with approaches for the realization of doping structures and nipi superlattices in GaAs by molecular beam exitaxy. In order to achieve a high degree of control of dopant incorporation during MBE, two processes must be understood and their relationship to substrate temperature and As_4 to Ga flux ratio must be determined. The first is the interaction behavior of the dopant with a GaAs surface; the second is the incorporation of the dopont into the growing film and its extent of electrical activity. Since there are several different factors which must be considered when producing alternating n- and p-type layers, we will describe some recent dopant incorporation results for GaAs films in rather more detail in section 3.2. Finally, in section 3.3 some doping structures in GaAs prepared by MBE will be presented.

2. ELECTRONIC PROPERTIES OF nipi CRYSTALS

2.1 Potential and electronic states.

In Fig. 1 a crystal with a "nipi" doping superstructure is shown schematically. The period d of this sequence of n-, p- and undoped (i= intrinsic) layers is the sum of the respective thicknesses of the layers, i.e.

$$d = d_n + d_p + 2d_i \tag{1}$$

The situation as shown represents an idealization. It should be noted that any system with a periodic doping profile which satisfies the requirement that the effective doping alternates between n- and p-type will show qualitatively a similar behavior as the idealized structures.

The range of values for the period d for which the most striking features of these systems are expected is by almost one order of magnitude larger than in the compositional superlattices. Such doping profiles with a width of a few 10^2 to 10^3 Å may be achieved by MBE, as described in section 3.

Apart from the period d the concentrations of donors and acceptors per doping layer, i.e. $n_D d_n$ and $n_A d_p$, respectively, are the important parameters which characterize the nipi crystal.

The excess electrons of all the donors can recombine with the holes of the acceptor atoms if $n_D d_n$ and $n_A d_p$ are equal and not too large. The resulting charge density per layer of the amount of

$$|\rho^{(2)}| = |e| n_D d_n = |e| n_A d_p \tag{2}$$

causes a periodic space charge field with maximum values of

$$F_{max} = 2\pi |\rho^{(2)}|/\varepsilon \tag{3}$$

and an amplitude of the space charge potential of the order of

$$V_o \simeq \pi |e\rho^{(2)}| d/\varepsilon$$

(e = elementary charge, ε = static dielectric constant of the semiconductor). With $n_D d_n = n_A d_p = 2 \cdot 10^{12}$ cm^{-2} and d= 10^3 Å, e.g., the values of the maximum field and

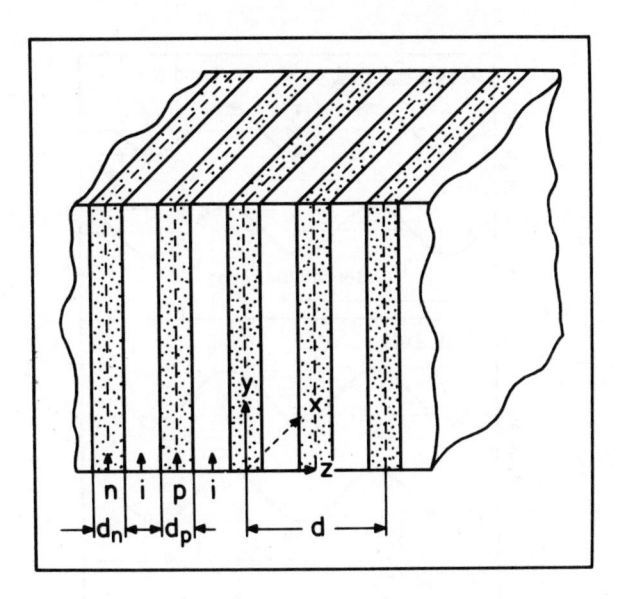

Fig. 1. Semiconductor crystal with periodic "nipi" doping superstructure of period d, schematically. The n- and p-layers contain $n_D d_n$ donors and $n_A d_p$ acceptors per unit cross section. The undoped (i=intrinsic) layers are not of substantial importance.

the potential amplitude are $F_{max} \approx 1.5 \cdot 10^5$ Vcm^{-1} and $2V_o \approx 0.75$ eV, respectively, for GaAs ($\varepsilon = 11$).

In Fig. 2 conduction and valence band edge, modulated by the space charge field, are shown schematically for various nipi crystals. Part a) corresponds to the situation just described with relatively small values of $n_D d_n = n_A d_p$. In this case one is dealing with a compensated nipi semiconductor with an effective band gap (defined as the energy difference between the lowest conduction subband and the uppermost valence subband edge)

$$E_{g,o}^{eff} \approx E_g - 2V_o \qquad (4)$$

In part b) of Fig. 2 the same value of $n_A d_p$ as before is assumed, $n_D d_n$, however, is increased. The potential amplitude is essentially the same as before, as the charge per layer cannot exceed the value of $|e| n_A d_p$. Also there are no holes in the p-layers. Each n-layer, however, contains

$$n_o^{(2)} = n_D d_n - n_A d_p \qquad (5)$$

electrons, as required by the condition of macroscopic neutrality, i.e. this crystal is n-type. Correspondingly the crystal is p-type if $n_A d_p > n_D d_n$.

Fig. 2c), finally, illustrates the situation for large values of $n_D d_n$ and $n_A d_p$. The superlattice potential amplitude has reached its maximum value of slightly more than one half of the energy gap E_g^o of the unmodulated semiconductor and the band gap has disappeared. There are electrons in the n-layers and holes in the p-layers in the ground state. Therefore, the nipi crystal resembles a semimetal.

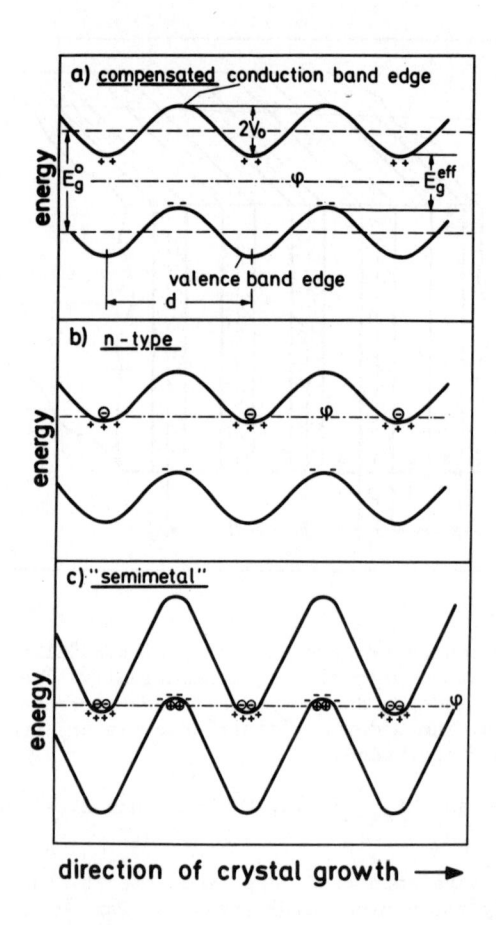

Fig. 2. Conduction and valence band edges with modulation by the periodic space charge potential. E_g^o = energy gap of the unmodulated semiconductor; E_g^{eff} = effective energy gap of the nipi crystal. + = ionized donors, − = ionized acceptors, ⊖ = electron, ⊕ = holes; ϕ = Fermi level. (a) compensated nipi crystal ($n_D d_n = n_A d_p$), (b) n − type nipi crystal ($n_D d_n > n_A d_p$), (c) heavily n- and p-doped nipi "semimetal".

A detailed knowledge of the exact shape of the superlattice potential and of the electronic subband structure is not required for the qualitative understanding of the electronic properties to be discussed below. A few remarks, however, should be made.

The shape of the space charge potential $v(z)$ can be easily calculated by integration of Poisson's equation for the intrinsic nipi crystal (Fig. 2a) for arbitrary doping profiles $n_D(z)$ and $n_A(z)$, if the potential fluctuations resulting from the statistical distribution of the dopants are neglected. For the calculation of the electronic states it is sufficient in most cases to solve Schrödinger's equation in the region around a single extremum of the superlattice potential only, as there is practically no propagation of the electronic states between adjacent layers, in contrast to the compositional superlattices with much smaller period d.

The electron states are characterized by their (2 dimensional) momentum $\vec{k}_{\parallel}$ for motion parallel to the layers with essentially the effective mass of the unperturbed semiconductor, whereas the motion perpendicular to the layers is quantized by the space charge field. The energy of electrons in the ν-th subband

$$\varepsilon_\nu(\vec{k}_\parallel) = E_\nu + \hbar^2 k_\parallel^2 / (2m^*) \qquad (6)$$

can be calculated from a simple one particle Schrödinger equation.[1]

The situation becomes more intricate with free carriers in the layers as depicted in Figs. 2b and c. In this case a selfconsistent solution is required which takes into account the partial screening of the fixed space charge by the other free **carriers** as well as exchange and correlation corrections. For a correct treatment one has to proceed in a way analogous to the solution of the 3-layer p-n^+-p system described in recent work by one of the present authors.[3,4]

The possibility of creating doping profiles of arbitrary shape allows to design quite different subband structures. For instance, one may realize systems where the subband spacing increases, decreases or stays constant with increasing subband index ν.

2.2 Transport and optical properties.

In this section the characteristic features of nipi crystals will be summarized. We will try to attribute equal importance to the scientific and to the device aspects as well.

1) The effective bandgap E_g^{eff} in nipi crystals is a quantity which may have any value between zero and E_g^o, the energy gap of the unmodulated bulk, depending on the choice of layer thickness and doping concentration as shown above. nipi crystals may be intrinsic, n-, or p-type, just as familiar semiconductors, but they also may have zero band gap and finite electron and hole concentration, similar to semimetals.

2) nipi crystals are extremely anisotropic with respect to their transport properties: The mobility of electrons and holes within the respective layers is high and does not differ strongly from a homogeneous semiconductor of the same doping concentration. The motion of carriers perpendicular to the layers, however, requires surmounting of the potential barrier of $2V_o$ by thermal excitation, e.g., provided that the doping period d is not small enough to allow for tunneling through the barrier. But even in the latter case the mobility perpendicular to the doping layers will be of the hopping type only and it will be small compared to the free motion parallel to the layers

3) Electrons and holes are separated from each other in real space (see Fig.3). The periodic space charge potential prevents electron hole recombination more or less perfectly. The probability for tunneling through the barrier may be small, but yet non-negligible for doping periods of a few 10^2 Å. The rate for tunneling recombinations, however decreases exponentially with the width and heights of the potential barrier. For periods d in the range of 10^3 Å only thermally activated recombination may be expected. Also the thermal recombination rate even at room temperature remains negligibly small for large and medium values of the superlattice potential amplitude, as it contains a Boltzmann factor $\exp(-2V_o/kT)$.

4) Another unusual property which is directly connected with the low recombination rates is the possibility of maintaining large deviations from thermal equilibrium as a quasistable state. Suppose, for instance, the nipi crystal has a concentration of n_o electrons and p_o holes in its ground state, in which the Fermi level for electrons and holes, by definition, is equal, i.e. $\phi_n^o = \phi_p^o$. For a negligibly

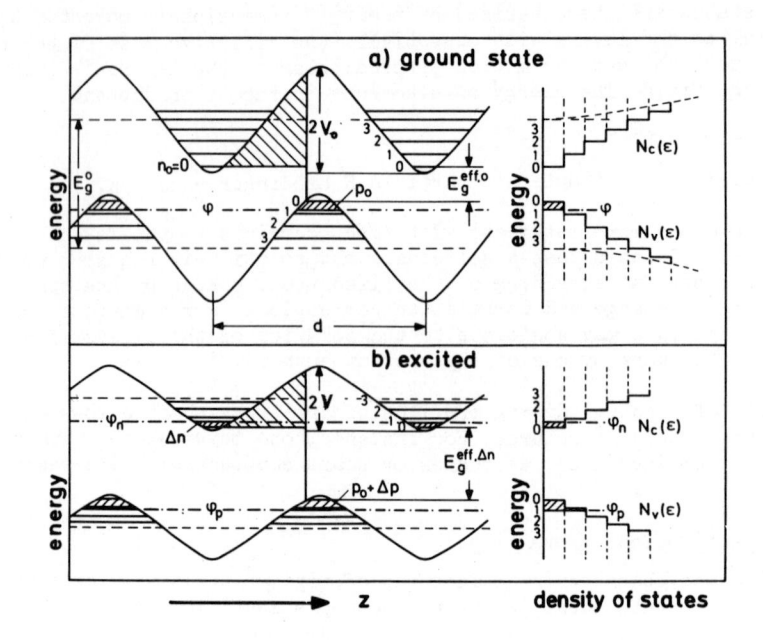

Fig. 3. nipi crystal (p-type) with subbands, schematically. The levels
$\nu = 0, 1, 2 \ldots$ correspond to the subband edges E_ν from equ. (6).
Shaded areas: tunneling barriers for electron hole recombination.
The density of states for the unmodulated semiconductor is depicted by dashed lines in part (a).
a) Ground state with $n_o = 0$, $p_o \neq 0$.
b) Excited state with $n = \Delta n$, $p = p_o + \Delta n$. The quasi Fermi level for electrons ϕ_n and for holes ϕ_p are different.

low recombination rate a situation with $n = n_o + \Delta n$ and $p = p_o + \Delta p$ and consequently with $\phi_n \neq \phi_p$ is quasistable. The condition of macroscopic neutrality requires $n = p$, just as in familiar semiconductors. It is interesting to note, that negative values for $n = p$ are also possible in nipi crystal. This situation corresponds to

$$n_p < n_o p_o$$

and
$$\phi_n - \phi_p < 0$$

for the hole bulk of the nipi crystal. In familiar semiconductor systems these conditions may be fulfilled only within a limited volume, namely in the depletion region near interfaces under reverse bias.

5) An increase of the electron and hole concentration by $\Delta n = \Delta p$ is directly correlated with a decrease ΔV_o of the amplitude of the periodic space charge potential. This relation follows from the fact that the amplitude of the space charge potential depends on the total charge per layer, i.e. the number of impurity atoms minus the number of carriers, which partly compensate the respective impurity charges. Equ. (2) has to be replaced by

$$|\rho^{(2)}| = |e| (n_D d_n - n^{(2)}) = |e| (n_A d_p - p^{(2)}) \tag{7}$$

For $n^{(2)} \neq 0$ and $p^{(2)} \neq 0$ the quasi Fermi levels for electrons and holes are rather close to the conduction band minima and valence band maxima, respectively (subband energies and the 2-dimensional Fermi energies $\varepsilon_p(n^{(2)}) = \hbar^2 \pi \, n^{(2)}/m_{eff}$ are typically of the order of 10 to 50 meV, only). Thus, it follows that the variation ΔV_o induced by changing the carrier concentration by $\Delta n = \Delta p$ is associated with a shift eU_{np} of the quasi Fermi levels with respect to each other, i.e., one has

$$eU_{np} = \phi_n(n_o + \Delta n) - \phi_p(p_o + \Delta p) \simeq 2 \Delta V_o$$

for a "nipi-semimetal", e.g., as depicted in Fig. 3b.

6) A change $\Delta n = \Delta p$ of the charge carrier concentration within the whole bulk of the nipi crystal can be achieved by lateral injecting contacts, as sketched in Fig. 4. The strongly doped n^+- (p^+-) zones represent electron (hole) injecting contacts with respect to the n-(p-) layers, whereas they form blocking p-n junctions with respect to the other type of layers.

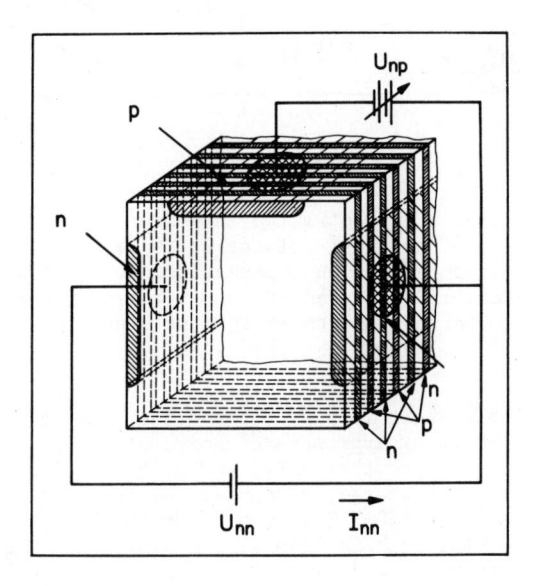

Fig. 4. nipi crystal with lateral electrodes and applied voltages U_{np} and U_{nn}. The n-doped contacts allow electron injection into the n-layers. They form blocking p-n junctions with respect to the p-layers. The effect of the p-contacts is analogous.

At this point possible device applications become evident: Fig. 5 shows that the charge carrier concentration of one type (of the electrons in this example) may be modulated between zero and large values by an external bias U_{np} applied between the n^+- and p^+- contacts if U_{np} is tuned from the threshold value U_{np}^{th} up to values close to E_g^o/e.

Simultaneously with the electron concentration also their conductivity σ_{nn} parallel to the n-layers is varied between zero and large values corresponding to homogenous n-crystals of comparable doping concentration. Thus, a current I_{nn} between the n-injecting electrodes as shown in Fig. 5 can be modulated without losses by variation of the bias U_{np}.

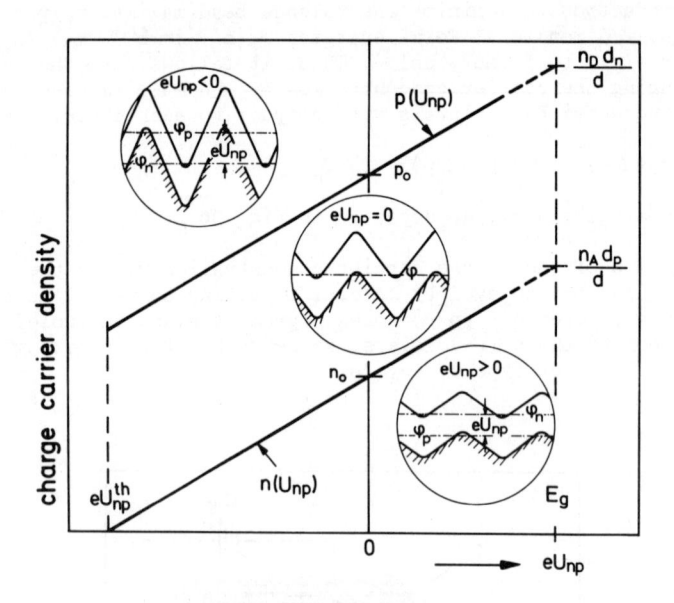

Fig. 5. Electron and hole concentration in a nipi "semimetal" with $p_o > n_o$ as a function of external voltage $U_{np} = (\phi_n - \phi_p)/e$, schematically. The electron concentration becomes zero at U_{np}^{th}. The insets shows the modulated band edges and the position of the quasi-Fermi levels for electrons and holes for the cases $eU_{np} \gtreqless 0$.

It is evident, that the threshold voltage U_{np}^{th} and the shape of $\sigma_{nn}(U_{np})$ may be "tailored" within wide limits by appropriate choice of d, $n_D d_n$, and $n_A d_p$, in particular if complicated doping profiles $n_D(z)$ and $n_A(z)$ are also considered. The arrangement just described, of course, resembles the familiar metal-oxide-semiconductor field effect transistor (MOSFET). Instead of the oxide semiconductor interface, however, the whole nipi bulk is involved in our case. The 3-dimensionality of the device permits high power modulation. Another advantage is the absence of any interfaces. This is particularly true in case of GaAs, where people working on the fabrication of MOS are faced to tremendous problems with interface states.[5] Other device applications such as capacitances with complicated C vs. U_{np} relations are obvious. Integrated circuits may be manufactured by means of an appropriate doping ion beam lithography.

Quite in analogy to the MOS devices, which have proved as an ideal tool for the study of the properties of 2-dimensional electron systems,[6] the same is expected for nipi crystals[1,2] and for 3-layer systems of the type p-n$^+$-p or n-p$^+$-n.[3,4,7] The carriers in these systems form a dynamically 2-dimensional many body system whose particle density and subband spacing may be tuned within wide limits. Apart from the structural differences (no interface, but homogeneous bulk materials, no change in dielectric constant, etc.) the doping system behaves differently in various respects:

The subband spacing decreases with increasing carrier concentration, in contrast to the MOS structures, which allows easily for population of higher subbands. There is also more flexibility to design the subband structure for some particular purpose by appropriate choice of the doping profile,[3,4,7] The fact that the nipi crystal is dynamically 2-dimensional but 3-dimensional with respect to configurational space has interesting implications for the plasmon spectrum, for instance. Differences resulting from interband transitions between electron and hole subbands will be discussed in connection with the optical properties.

7) The optical properties of a semiconductor are also drastically modified by a nipi superstructure. This applies to the absorption and the emission of light, as well.

a) Due to the reduction of the effective energy gap in nipi crystals the absorption of photons of the energy

$$\hbar\omega > E_g^{eff} \simeq E_g^o - 2V \tag{8}$$

is energetically allowed. The absorption coefficient $\alpha(\omega)$, however, may be quite low near the threshold energy, if the tunneling barrier entering into this process is high and rather wide (see Fig. 3). But with increasing photon energy $\alpha(\omega)$ becomes large quite below the gap E_g^o of the unmodulated semiconductor, even at doping periods of a few 10^3 Å.

This absorption tail for $\hbar\omega < E_g^o$ can be interpreted as Franz-Keldysh effect[8] due to the internal space charge fields. The absorption coefficient of a nipi crystal may be strongly modulated by rather small variations of the internal field. Such changes of the internal fields may be self-induced by the electrons and holes generated in the absorption process or the modulation may be achieved by injection or extraction of charge carriers via a varying external voltage U_{np} applied to lateral electrodes, as described before.

Another absorption related property of particular interest is the photovoltaic effect of nipi crystals. The electron hole pairs are separated by the space charge fields which are present in the whole crystal and there upon they are collected in the respective layers. The photovoltaic efficiency is expected to be very high, as the active zone is not just a layer of finite thickness (the minority carrier diffusion length in p-n solar cells, e.g.), but the whole nipi bulk. In addition to that, the efficiency is increased by the extended spectral range for absorption compared with the unmodulated bulk.

b) Transitions between the lowest conduction and the uppermost valence subbands require tunneling of the carriers through a potential barrier (see Fig.3). Therefore, the intensity of luminescent electron hole recombination processes does not only depend on the concentrations n and p of electrons and holes, but it depends exponentially on the heights ($\simeq 2V$) and the width ($\simeq d/2$, possibly considerably less) of this barrier. The frequency of the photons emitted, given by

$$\hbar\omega < \phi_n - \phi_p, \tag{9}$$

corresponds to the external voltage $eU_{np} = \phi_n - \phi_p$ in the case of electroluminescence. The optimum working point for such a tunable light source may be adjusted to the actual requirements by appropriate choice of the doping profiles. The population inversion and, hence, the active zone extends over the whole nipi crystal in contrast to light emitting diodes. The reabsorption for photons with $\hbar\omega < eU_{np}$ should be very weak. Therefore, it is expected that the threshold current of a tunable nipi-injection laser should be low and that the lifetime may be long because of the low local power densities required.

Apart from the optical properties just mentioned above, which may become important with respect to device application, there are again features of more basic interest related to the interaction between two 2-dimensional many body systems.

In an idealized one-particle picture (no potential fluctuations due to random impurity distribution) the luminescence spectrum should reflect the joint density of states distribution of conduction and valence subbands. The observed emission spectrum will show how far the momentum conservation condition is relaxed by free carrier-impurity and by free carrier-free carrier scattering. The emission spectrum also may reveal deviations from the one-particle picture due to retardation effects in the many body system.

A number of features, which are specific for nipi crystals, have even not been stretched in this review for the sake of brevity. The possibility of high frequency modulation of light absorption and emission via sandwich like electrodes (the response time of the system is limited by the electrode capacity only), e.g.

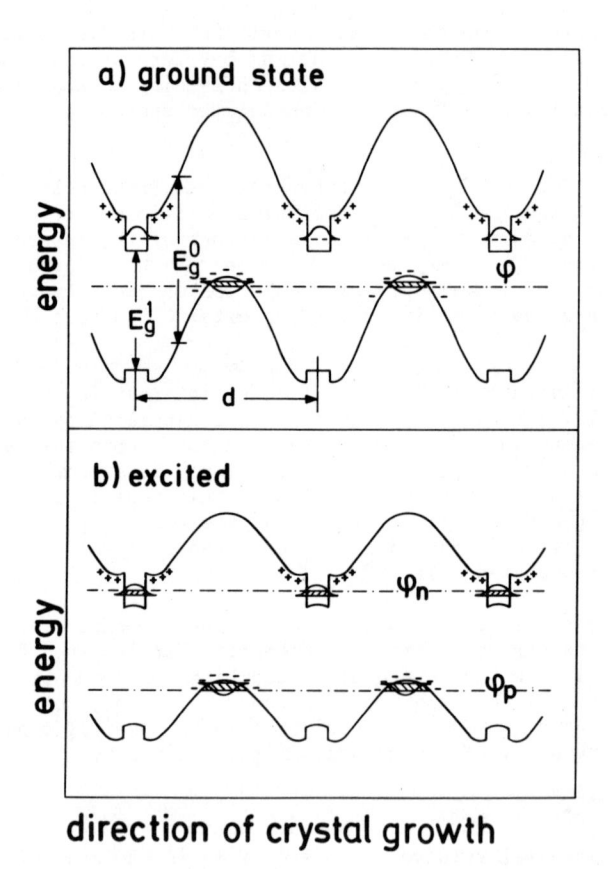

Fig. 6. Doping superlattice, modified by the incorporation of a layer of undoped lower band gap material (E_g^1). The system behaves qualitatively like a normal nipi crystal. The mobility in the n-layers, however, is much higher due to the local separation between electron and scattering centers (i.e. the randomly distributed ionized donors). a) p-type crystal in the ground state (p_o; $n_o = 0$). b) excited state ($p = p_o + \Delta p$; $n = \Delta n = \Delta p$).

has not been discussed. Instead of seeking for completness, however, it appeared
more appropriate at the present state of the art to exemplify in which respects
doping superlattices represent a new material with intriguing properties from the
aspect of scientific research and of device application as well.

The discussion of electronic properties of nipi crystals may be concluded by an
outlook on more sophisticated structures. In Fig. 6 a nipi crystal is shown which
is modified by the incorporation of an undoped layer of lower band gap material
(GaAs in a $Ga_{1-x}Al_xAs$ crystal, or Ge in a GaAs crystal, e.g.) in the n-region. The
electronic properties of such a system do not differ qualitatively from the normal
nipi crystal. The mobility of the electrons, however, will be much higher, since
the randomly distributed impurity potentials are centered in a region where the
electronic wave function amplitude is already quite small. The drastic increase
of mobility achieved by such an arrangement has been demonstrated recently for the
compositional superlattice by the work of Dingle, Störmer, Gossard and Wiegmann.[9]

3. MBE GROWTH OF DOPING STRUCTURES IN GaAs

3.1. MBE Growth Process

The technique of molecular beam epitaxy (MBE) is based on the condensation of directed thermal beams of atoms or molecules of the constituent elements on a heated
substrate under ultra-high vacuum conditions. In the preceeding article it has been
pointed out that the most distinguishing characteristics of MBE compared to the
more established methods such as liquid phase epitaxy (LPE) and vapor phase epitaxy (VPE) are the low growth temperature, the low growth rate, the ability to
stop or initiate growth abruptly, and the smoothing of the substrate surface. As
a result of all these favourable features the MBE technique provides a precise control over composition, thickness, and doping profile in the direction of growth
on an atomic scale. In addition, an excellent uniformity of the crystal properties
over the whole area of the substrate can be achieved.

Joyce and Neave have found[10] that the electrical and optical properties of GaAs
films deteriorate when grown below $400°C$. The high temperature limit for MBE growth
is given by the congruent evaporation of GaAs which occurs at about $625°C$[11]. Thus,
for practical applications, GaAs films can be grown in the temperature range of
400 to $620°C$. In order to ensure a stoichiometric growth, an excess flux of arsenic must always be supplied. Usually the GaAs films are grown with the arsenic
stabilized (2 x 4) surface structure because the so required growth conditions
are more easily maintained.

Starting from elemental As_4 as arsenic source, the as-grown GaAs films are usually
found to be p-type. The hole concentration ranges from 1×10^{15} to about $6\times10^{14} cm^{-3}$.
The residual acceptor is probably carbon. It is expected that the impurity level
can be reduced by about half an order of magnitude if a sample load-lock system
is used.

Starting from GaAs as arsenic source, the as-grown films are always unintentionally doped with Si from the As source, leading to n-type films in the low $10^{14} cm^{-3}$
range with room temperature mobilities between 6000 and 7000 $cm^2/Vsec$.

The incorporation of a controlled amount of electrically active impurities into
the growing GaAs film has been achieved by using additional effusion cells, which
contain the appropriate dopant elements. The ultra-high vacuum system which is
presently being used for MBE growth of doping structures in GaAs is of the vertical evaporation type and is schematically shown in Fig. 7. In addition to the
standard techniques of Auger Electron Spectroscopy (AES) and Reflection High Energy

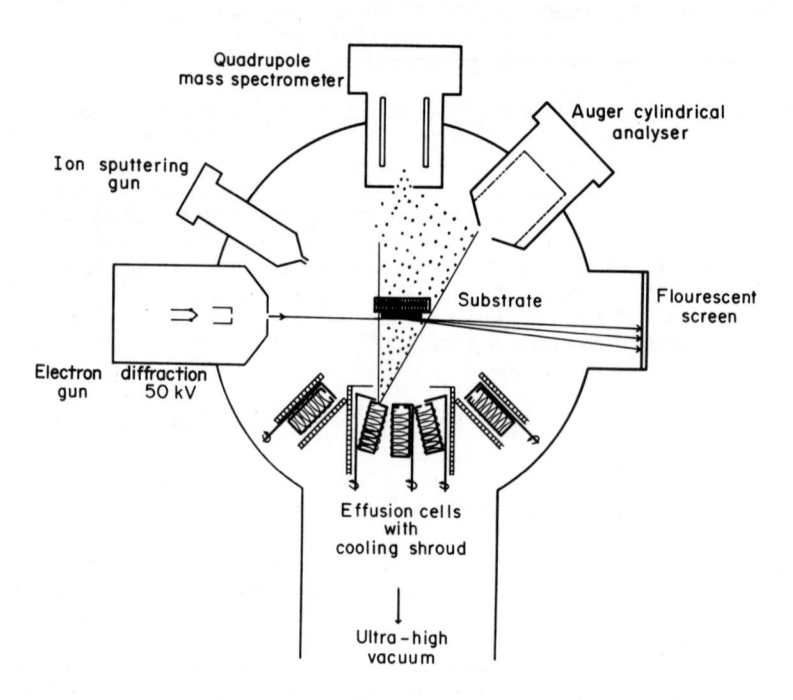

Fig. 7. Schematic diagram of the experimental arrangement for growth of doping structures in GaAs by molecular beam epitaxy.

Electron Diffraction (RHEED), we have incorporated the components for Secondary Ion Mass Spectroscopy (SIMS) into our MBE growth apparatus[12]. SIMS is a sensitive analytical tool to detect even trace impurities (intentional and unintentional ones) in the substrate and in the as-grown material and to perform depth profiling of multilayer structures.

Each dopant source is provided with an externally controlled individual shutter. Operation of these shutters permits rapid changing of the beam species in order to alter abruptly the dopant concentration in the growing film. As the growth rate during MBE is low, about 0.5 µm/h, the shutter time is much less than the time for the growth of a monolayer. Thus, abrupt profiles can be realized, since diffusion and surface accumulation are normally negligible at the low epitaxial growth temperature for MBE.

3.2. Dopant Incorporation

Growth and doping of GaAs by MBE involves processes which run far from equilibrium. Therefore, kinetic effects are dominant, and the doping process is strongly dependent on the flux of the impurity atoms reaching the growth interface, on their sticking probabilities, and on their surface lifetimes. The mechanism of dopant incorporation depends on the species. It is well understood only for a few elements, and our understanding of the general doping process is still largely on an empirical level. A number of suitable donor and acceptor impurities for MBE GaAs along with some of their characteristic properties are summarized in Tables 1 and 2, resp. In practice all elements with the exception of zinc, which have been used successfully as dopants, have unity sticking coefficients over the whole range of growth conditions. This has been established by modulated molecular beam experiments[13].

Table 1. Promising n-Type Dopants in MBE-GaAs

Element	Incorporation Behaviour	Degree of Compensation	Maximum Achievable Electron Concentration	Remarks
Si	predominantly donor	fairly low	5×10^{18} cm^{-3}	evaporation temperature fairly high ($\sim 1150°C$)
Ge	depends strongly on growth conditions	can be high	4×10^{18} cm^{-3}	donor at high As$_4$ to Ga flux ratio and low growth temperature
Sn	donor only	low	8×10^{18} cm^{-3}	tends to accumulate at the growth surface

3.2.1. <u>Donor Impurities in MBE GaAs</u>. The most convenient n-type dopant for GaAs in any epitaxial growth technique is tin. Contrary to the other group IV elements Si and Ge, which act as amphoteric impurities in GaAs and can form donor or acceptors depending on the reaction conditions, with tin only n-type GaAs can be grown by LPE, VPE, or MBE. Sn doped GaAs films prepared by MBE exhibit a low degree of compensation. High carrier concentrations with excellent mobilities can be readily achieved, and the photoluminescence intensity for Sn-doped MBE GaAs layers was found to be greater than for high-quality Sn-doped LPE GaAs layers[14]. The only major disadvantage of using tin is its tendency to accumulate at the GaAs surface during epitaxial growth[15]. The mechanism of tin incorporation into MBE GaAs is now well established, and its enrichment at the growth surface can be readily explained as due to a surface rate limitation to its incorporation[16]. As a result, no abrupt doping profiles perpendicular to the growing surface can be achieved, if tin is used as donor impurity in MBE GaAs.

Very abrupt doping profiles in MBE GaAs can be obtained with Ge and Si as donor which are both amphoteric, as first observed by Cho and Hayashi[17]. This means that the site incorporation of the impurity elements depends critically on the As$_4$ to Ga flux ratio and on the substrate temperature. Thus, p/n junctions in GaAs can be grown by simply changing the growth conditions in an appropriate manner[17]. Under As-stable conditions with a (2x4) surface reconstruction Ge and Si are predominantly incorporated as donor. Cho[18] has demonstrated that a periodic doping profile of donors in MBE GaAs is only well resolved when the film is doped with Ge or Si, respectively. This is illustrated for Ge doping in Fig. 8. The square trace has been calculated from the dopant beam intensity. The ultimate sharpness of the measured profile in C-V experiments, of course, is limited by the majority-carrier diffusion, and therefore is broadened even for an abrupt dopant profile[19].

Until recently, Ge has been found to be not particularly useful as a practical n-type dopant in MBE GaAs, because it had exhibited a high degree of compensation when the films were grown under standard As-stable growth conditions[17]. We were now able to show that relatively uncompensated n-type films may be deposited with sufficiently large As$_4$ fluxes relative to Ga fluxes ($\sim 10/1$) and with low substrate

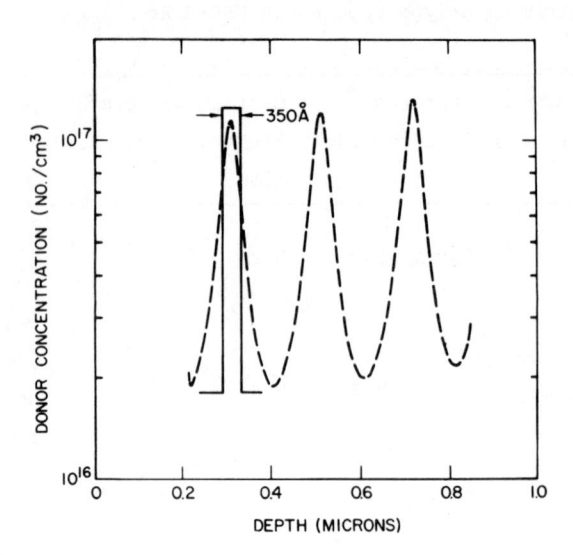

Fig. 8. Periodic doping profile of Ge-doped GaAs grown under As-rich conditions (from reference 18). The solid line has been derived from the dopant beam intensity; the dashed line represents the measured profile.

temperatures (500°C)[20]. Under these conditions sufficient Ga vacancies can be created and Ge is predominantly incorporated as a donor. As a result, values of N_D/N_A up to 4/1 have been obtained.

As a result, we have the choice between only three suitable donor impurities for MBE of GaAs, but none is without unfavourable attributes. For the fabrication of doping structures an additional striking feature must be considered. Ge and to a lesser extent Si have a tendency to exhibit a memory effect. This means that these elements are always present as residual impurities in the grown film even after several subsequent runs without intential Ge and Si doping, respectively. We attribute the presence of Si or Ge, resp. to contamination with volatile Ge and Si oxide species resulting from previous evaporation of these elements in the system.

3.2.2. <u>Acceptor Impurities for MBE GaAs.</u> By now four suitable acceptor impurities for MBE GaAs have been found (Table 2). The most recently established one is beryllium, which behaves as an almost ideal p-type dopant in MBE GaAs[21]. There are no complications during incorporation of Be resulting from a low sticking coefficient or from diffusion and segregation effects. The calculated and measured values of free acceptor concentrations are directly proportional to the Be effusion cell temperature (i.e. Be vapour pressure). Each incident Be atom produces one free acceptor, and it provides a fairly shallow acceptor level only 30 meV above the valence band edge. Ilegems has shown[21] that measured doping profiles follow very closely ideal profiles calculated from the dopant beam intensity. Fig. 9 shows a typical dopant profile obtained from a n^+-p-p^+ diode structure in which the p-layer consisted of a series of sharp Be dopant pulses superimposed on a uniform Be dopant background. The measured "as-grown" profile nearly exactly matches the theoretical free-carrier profile[19,30] (top solid curve). Little change was observed even after annealing for 1 h at 800°C under $\sim$10 atm As_4 pressure. More recently, McLevige et al. have shown[22] the diffusion coefficient of Be at 900°C in Be-doped

Table 2. Promising p-type Dopants in MBE-GaAs.

Element	Incorporation Behaviour	Degree of Compensation	Maximum Achievable Hole Concentration	Remarks
Be	acceptor only	low	5×10^{19} cm^{-3}	vapour is poisonous
Mn	acceptor only	low	1×10^{18} cm^{-3}	film perfection deteriorates at $p > 10^{18}$ cm^{-3}, deep acceptor level
Ge	depends strongly growth conditions	fairly low	5×10^{18} cm^{-3}	acceptor at low As$_4$ to Ga ratio and high growth temperature
Zn$^+$	acceptor only	low	2×10^{19} cm^{-3}	sophisticated technique required for low energy ion production

MBE GaAs $[(0.5 \text{ to } 1) \times 10^{-10} \text{cm}^2/\text{sec}]$ to be two orders of magnitude lower than that observed for implanted Be in GaAs of equal concentration of $(2 \text{ to } 3) \times 10^{19}$ cm^{-3}. Beryllium is, therefore, expected to be a very useful acceptor in MBE growth of GaAs.

Before Be has been established to be a suitable acceptor, manganese was widely used for MBE growth of GaAs under As-rich growth conditions[23]. However, the strong interaction of Mn with Ga during growth results in a deterioration of film perfection at doping levels 10^{18} acceptors per cm^3. Furthermore, Mn forms a rather deep acceptor (113 meV above the valence band edge) and is thus not fully ionized at room temperature.

When the surface of the growing GaAs film is Ga-stabilized, Ge is incorporated predominantly as acceptor on As sites, and thus p-doped films result. The major problem for MBE growth of p-type GaAs films doped with Ge, which exhibit a low level of compensation and reasonable mobilities, has been the very small stability range of the required Ga-stabilized growth conditions with respect to the Ga flux while all other conditions remain constant. For practical purposes it has not been possible to maintain a Ga-stabilized (4x2) surface structure during continuous growth, resulting in a smooth mirrorlike surface appearance similar to that achieved routinely under As-rich growth conditions. Only a small increase (∼ 3%) in the Ga flux above the minimum value required to produce the Ga-stable (4x2) reconstruction resulted in the formation of free gallium on the surface. Thus, until recently, most of the electrical and optical characterizations of MBE grown p-type GaAs films doped with Ge have been performed on samples with rough surfaces (see for example reference[21]).

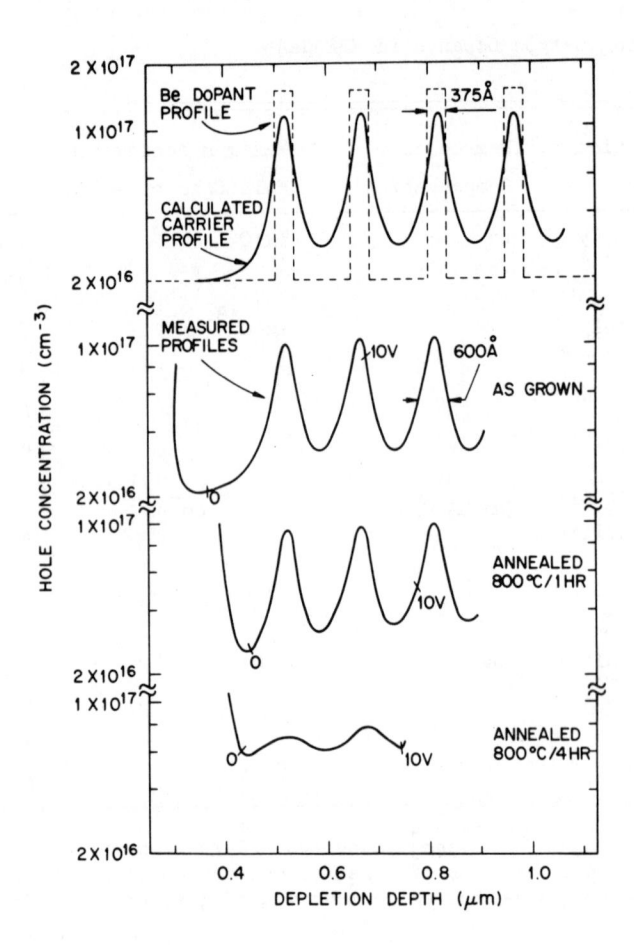

Fig. 9. Dopant profile obtained from a n^+-p-p^+-GaAs diode structure with periodic Be dopant modulation in the p-region (from reference 21). The dashed line on the top represents the arriving dopant profile. The top solid line represents the calculated theoretical free-carrier profile[19,30].

We have now established improved reaction conditions which enable the growth of Ge doped p-type GaAs films with extremely smooth surfaces[20]. The concentration of arsenic vacancies on the growth surface depends on the incident As_4 to Ga flux ratio as well as on the substrate temperature[24,25]. Thus, sufficient As vacancies can also be created, if the substrate temperature is increased to $T_S > 600°C$, while the As_4 to Ga flux ratio is kept slightly above the value required to maintain the As-stable (2x4) reconstruction at 550°C (i.e. $J_{As_4}/J_{Ga} \approx 2/1$). Under these growth conditions Ge is principally incorporated as acceptor without an onset of 3-dimensional growth, and the resulting p-type GaAs films exhibit a reasonable low level of compensation. The elevation of the substrate temperature from $T_S \sim 550°C$ to $T_S > 600°C$ is accompanied by an overall change of the As-stable (2x4) surface structure to a (1x1) structure without any reconstruction. The conversion of the surface structure can so be used to monitor the site incorporation of the incident Ge atoms in the RHEED pattern.

The ability to produce smooth GaAs films with abrupt changes in the carrier type, which are relatively uncompensated and free of lattice strain, make Ge a very promising dopant for the fabrication of p/n junctions in GaAs.

Neutral zinc shows a very complex surface interaction behaviour on (100) GaAs and its sticking coefficient in zero at substrate temperatures $\geq 100°C$ [26]. Therefore, ionized atoms have to be used for an effective acceptor incorporation during MBE growth as indicated in Table 2. Naganuma and Takahashi[27] have shown that an effective sticking coefficient of 0.01-0.03 can be achieved by using low energy (0.2 - 1.5 keV) Zn^+ ions. The film have hole mobilities comparable to those of VPE and LPE material, but there are still some problems with the minority-carrier properties.

In the future, the use of low energy ion beams might provide a very attractive doping technique in order to write in-situ 3-dimensional doping structures in GaAs during epitaxial growth.

3.3. MBE Grown Doping Structures in GaAs

The fabrication of GaAs IMPATT[28] and Varactor diodes[29] has provided excellent examples of the capability of MBE to produce both abrupt and precisely tailored donor impurity profiles perpendicular to the growing surface. The low-high-low profile in the active layer of an IMPATT diode with improved DC to RF conversion efficiency is schematically shown in Fig. 10[28]. A narrow avalanche zone is confined between the metal Schottky barrier and an extremely thin, highly doped n layer. The structure is completed by a moderately doped n-type drift region and an n^+ buffer layer. This structure clearly represents a severe test of doping and thickness control achieved by MBE.

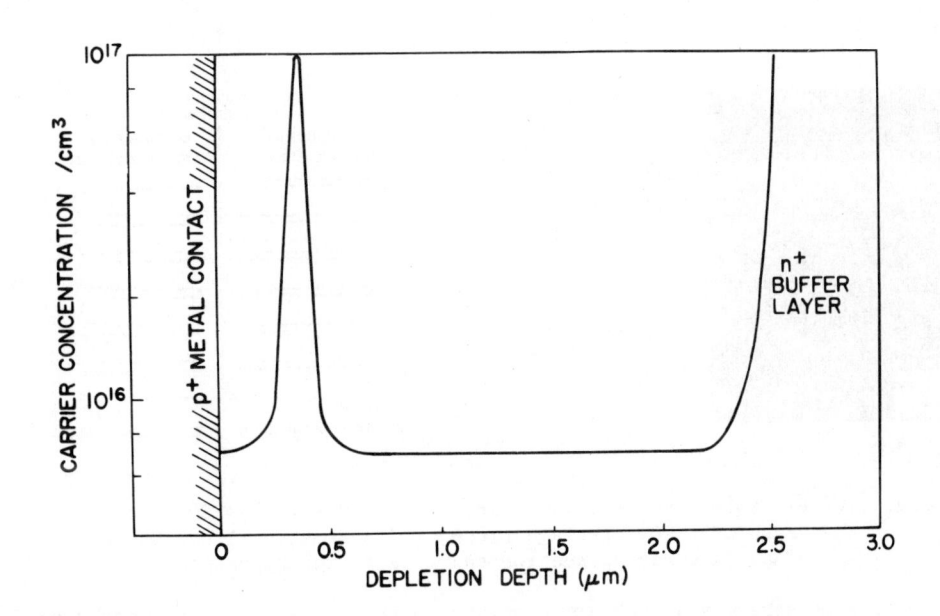

Fig. 10. Typical carrier profile of a low-high-low IMPATT diode with improved DC to RF conversion efficiency (from reference 28).

For 3 W output at an operating frequency of 11.7 GHz the efficiency of diodes fabricated from the shown low-high-low structure was 18% and the noise measure only 44 dB[28]. The uniformity of growth across the substrate area which can be achieved with MBE resulted in very similar electrical parameters for all of the IMPATT's fabricated frome one wafer; the standard deviation in the reverse breakdown voltage (median value 45.3 V) for 31 randomly selected diodes was only one volt.

The efficiency and power capabilities of the low-high-low diodes are higher than those obtained with flat doping-profile diodes, but without an increase in noise measure. Therefore, the performance so far achieved makes them quite attractive for microwave power amplification application.

With the evaluation of suitable acceptor and donor impurities in section 3.2 we have provided the most important prerequisite for the fabrication of p/n junctions, p/n multilayers, and nipi superlattices in GaAs by molecular beam epitaxy. In practice there are two methods to accomplish abrupt changes in the carrier type during MBE growth;

(1) in addition to the Ga and As_4 effusion cells two sources containing different doping elements (e.g. Be for acceptor and Si for donor) are used, and the shutters of the dopant cells are opened and closed in a precisely controlled manner.

(ii) only a single source containing an amphoteric doping element (e.g. Ge) is used in addition to the Ga and As_4 cells. The dopant incorporation on either Ga sites ($\rightarrow$ n-type layer) or As sites ($\rightarrow$ p-type layer), regulated by the As_4 to Ga flux ratio and substrate temperature, is monitored via the abrupt change of surface reconstruction in the RHEED pattern.

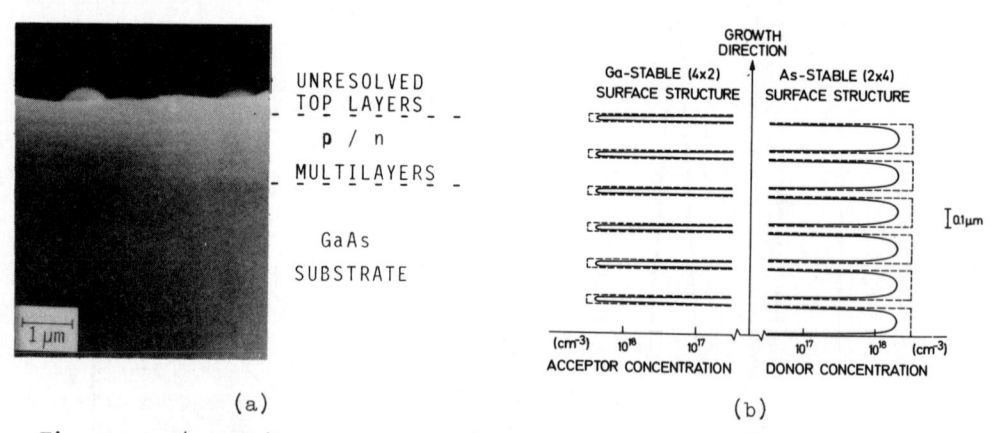

Fig. 11. p/n multilayer structure in Ge-doped GaAs; (a) SEM micrograph of (110) cleavage plane; (b) schematic illustration of surface reconstruction and Ge site incorporation.

Because of some unresolved problems associated with the memory effect observed during doping with silicon, we have been using the second method with an amphoteric dopant source to grow periodic structures with uniform dopant concentration within each individual layer and controlled thickness with constant periodicity.

As an example a scanning electron micrograph of the (110) cleavage plane of a p/n GaAs multilayer structure doped with Ge only is shown in Fig.11(a). Twelve alternating layers are clearly resolved in the shown structure. Each individual p-type layer is 0.04 μm thick and has a room temperature hole concentration of 2.5×10^{18} cm^{-3}.

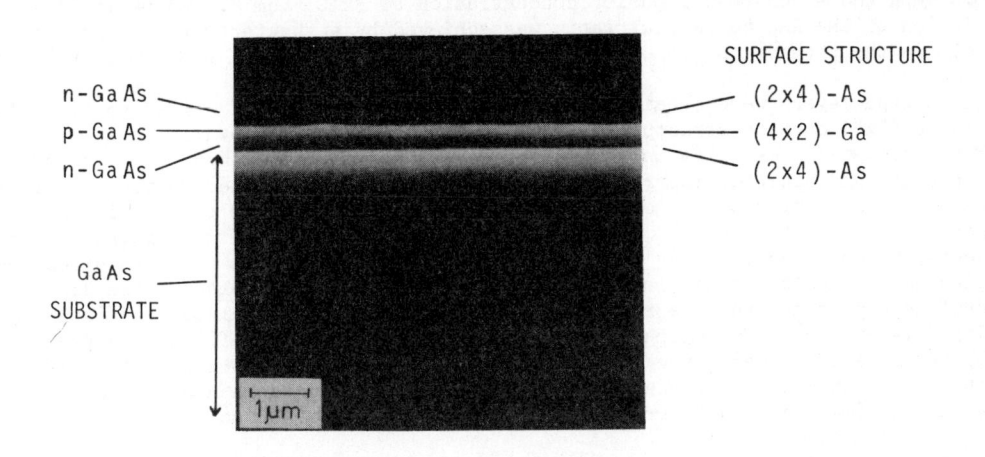

Fig. 12. 400 Å thick p-GaAs layer (p=2.5×10^{18} cm^{-3}) sandwiched between two n-GaAs layers (n=2×10^{18} cm^{-3}). SEM micrograph of (110) cleavage plane.

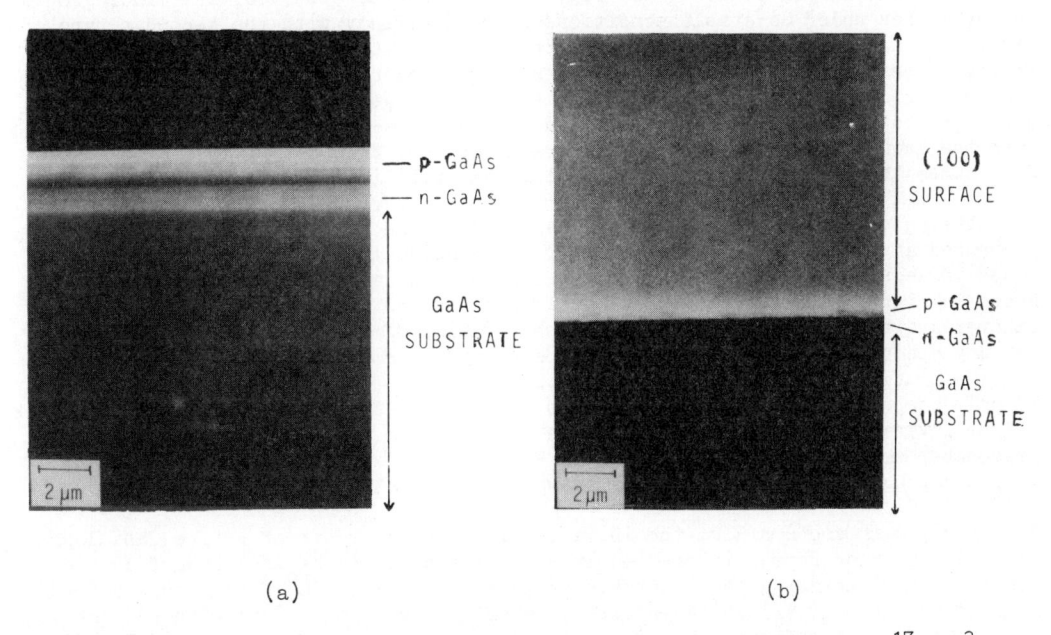

(a)   (b)

Fig. 13. SEM micrograph of p/n junction in Ge-doped GaAs (p=n=3×10^{17} cm^{-3}) grown on n$^+$-GaAs substrate; (a) (110) cleavage plane; (b) specimen titled at 45° with respect to the incident electron beam showing the smooth surface of the top p-type layer.

Each n-type layer has a thickness of 0.16 μm and an electron concentration of 2×10^{18} cm^{-3} at room temperature. The structure was obtained by impinging molecular beams containing As_4, Ga and Ge simultaneously on the substrate and by changing periodically the As_4 to Ga ratio at a constant substrate temperature of $530°C$. Both the Ga and Ge flux were kept constant resulting in a constant growth rate of 0.4 µm/h and a constant impurity concentration of 3×10^{18} cm^{-3}. The periodical variation of the As_4 to Ga flux ratio was achieved by a shutter with a 1 mm central hole in the path of the As_4 beam, and it was monitored in the RHEED pattern.

The n-type layers were obtained at an As_4 to Ga flux ratio $\geq 2/1$ leading to an As-stable (2x4) surface structure, and the p-type films were grown at an As_4 to Ga flux ratio $< 1/1$ resulting in a Ga-stable (4x2) surface structure. The widths of the donor and acceptor incorporation were accurately determined by the time interval between turning on and off the aperture into the As_4 beam path. This is schematically indicated in Fig. 11(b). The dashed line profile in this diagram represents the arriving impurity profile incorporated either on Ga or on As sites depending on the As_4 to Ga flux ratio. The solid line superimposed on the dopant profile represents[19,20] the theoretical free-carrier profile calculated for isolated doped layers. The slope of the impurity distribution profile was determined by the rate of changing the As_4 to Ga flux ratio from (2x4)As-stable to (4x2) Ga-stable growth and vice versa. For the shown structure this change was achieved within a time required to grow less than 5 monolayers (~10 sec.).

On top of the 12 clearly observable alternate p/n-GaAs layers additional 4 p/n layers were deposited in the same growth run which, however, are no longer resolved due to the increasingly unstable growth conditions during growth of the p-type layers (c.f. section 3.2). After an overall of 16 layers growth had to be stopped because of the onset of 3-dimensional growth of Ga droplets. From the cross-section of the p/n multilayer structure in Fig. 11(a) it can be observed that minor growth instabilities under Ga-stable conditions started already with the second p-type layer showing a slightly wave-like growth surface. With increasing number of the p-type layers the overall surface becomes rough, because the surface roughness induced during growth of the p-type films can no longer be balanced by the subsequent growth of a thicker n-type film under stable As-rich growth conditions. Even the growth of the shown structure required very thin p-type layers followed by 4 times thicker n-type layers for a limited stabilization.

The obtained wave-like bending of the majority of the individual layers has by now prevented an experimental observation of 2-dimensional electronic properties proposed in section 2 in our doping multilayer structures. However, the structure shown in Fig. 11(a) is a convenient structure for the demonstration that such multilayer growth is feasible and for the observation of the individual layers without any special etching in the scanning electron microscope.

If only one, rather thin (400 Å), p-type GaAs layer is sandwiched between two thick n-type layers equally doped with Ge, the interface to the final n-layer remains reasonably smooth. This is indicated in Fig. 12, where the (110) cleavage plane of a GaAs doping sandwich structure is shown.

As described in section 3.2, the instabilities during growth of p-type GaAs doped with Ge can be overcome, if the required As vacancies on the growth surface are generated by increasing the substrate temperature to $T_G \geq 600°C$ at a constant As_4 to Ga flux ratio of 2/1 which results in stable growth conditions with respect to the Ga flux. Furthermore, a reduction of the substrate temperature to $T_S \leq 500°C$ during generation of Ga vacancies under As-rich growth conditions yields n-type films which exhibit a rather low level of compensation and improved electron mobilities. We have now been using these modified growth conditions routinely to grow improved[20] p/n junctions and p/n multilayers in Ge-doped GaAs by molecular beam epitaxy. In Fig. 13(a) the (110) cleavage plane of a p/n junction in GaAs thus grown on n^+ substrate at a growth rate of 0.65 µm/h is shown. Also shown is the surface of the top p-type layer of this diode, the appearance of which is perfectly smooth (Fig. 13(b)). Finally Fig. 14(a) shows the current-voltage characteristics of two representative diodes at room temperature. The reverse breakdown voltage of the dio-

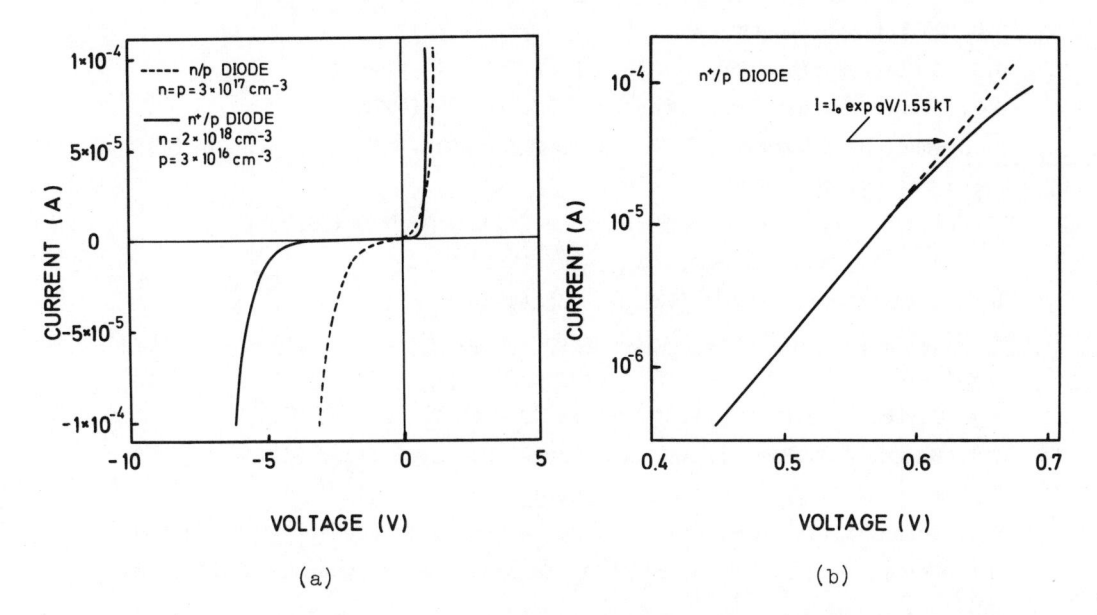

Fig. 14. Junction behavior of two representative Ge-doped MBE GaAs diodes at 297 K; (a) current-voltage characteristics; (b) exponential current-voltage dependence in the forward direction.

des is well below the values expected for bulk material[31]. This can be attributed to surface effects. In the forward direction at voltages above 0.4 V the current-voltage characteristics can be fitted by the equation $I = I_o \exp qV/1.55 kT$ (Fig. 14(b)), which is expected if diffusion and recombination mechanisms are both comparable for the current across the p/n junction[31].

In the experiments described above, the change in doping between layers was controlled by the substrate temperature and monitored in the RHEED pattern. Therefore, the interface between the layers was well defined and smooth. Since the growth rate during MBE growth is in the order of angstroms per second, the thickness of each layer can be precisely controlled. The uniform doping profile in each individual layer only requires constant substrate and effusion cell temperatures which are now being monitored by a computer controlled temperature regulation system[32]. The addition of a second Ge source enables us to realize instantaneous changes in the impurity level by opening and closing the appropriate shutters in coincidence with the changes of the substrate temperature. Since the atomic size of Ge is very similar to that of Ga and As, the Ge-doped GaAs films are relatively free from lattice strain. As a result, Ge has turned out to be a very promising dopent for the fabrication of p/n multilayers and p/n superlattices in GaAs by molecular beam epitaxy.

Acknowledgment

The authors would like to thank Miss H. Willerscheid and Mr. A. Fischer for technical assistance, and Miss G. Keck for typing the cameraready manuscript. The experimental work was sponsored by the Bundesministerium für Forschung und Technologie (contract No. NT 0785 5).

REFERENCES

1. G.H. Döhler, phys. stat. sol. (b) 52, 79 (1972).
2. G.H. Döhler, phys. stat. sol. (b) 52, 533 (1972).
3. G.H. Döhler, Surface Science 73, 97 (1978).
4. G.H. Döhler, Verhandlungen der DPG VI., 13, 89 (1978).
5. M. Hirose, S. Yokoyama and Y. Osaka, phys. stat. sol. (a) 42, 483 (1977).
6. For a review, see: Surface Science 73, (1978).
7. I. Uri and J.W. Holm-Kennedy, Surface Science 73, 179 (1978).
8. W. Franz, Z. Naturforschung 13a, 484 (1958);
   L.V. Keldysh, Sov. Phys. JETP 34, 788 (1958).
9. R. Dingle, H.L. Störmer, A.C. Gossard and W. Wiegmann, Appl. Phys. Letts. 33, 665 (1978).
10. B.A. Joyce and J.H. Neave, J. Crystal Growth 43, 204 (1978).
11. C.T. Foxon, J.A. Harvey, and B.A. Joyce, J. Phys. Chem. Solids 34, 1693 (1973).
12. K. Ploog and A. Fischer, Appl. Phys. 13, 111 (1977).
13. C.T. Foxon, M.R. Boudry, and B.A. Joyce, Surface Science 44, 69 (1974).
14. H.C. Casey, Jr., A.Y. Cho, and P.A. Barnes, IEEE J. Quantum Electronics QE-11, 467 (1975).
15. K. Ploog and A. Fischer, J. Vac. Sci. Technol. 15, 255 (1978).
16. C.E.C. Wood and B.A. Joyce, J. Appl.Phys. 49, 4854 (1978).
17. A.Y. Cho and I. Hayashi, J. Appl. Phys. 42, 4422 (1971).
18. A.Y. Cho, J. Appl. Phys. 46, 1733 (1975).
19. D.P. Kennedy, P.C. Murley, and W. Kleinfelder, IBM J. Res. Dev. 12, 399 (1968).
20. K. Ploog, A. Fischer, and H. Künzel, Appl. Phys., to be published.
21. M. Ilegems, J. Appl. Phys. 48, 1278 (1977).
22. M. V. McLevige, K.V. Vaidyanathan, B.G. Streetman, M. Ilegems, J. Comas, and L. Plew, Appl. Phys. Lett. 33, 127 (1978).
23. M. Ilegems, R. Dingle, and L.W. Rupp, Jr., J. Appl. Phys. 46, 3059 (1975).
24. A.Y. Cho, J. Appl. Phys. 47, 2841 (1976).
25. J.H. Neave and B.A. Joyce, J. Chystal Growth 44, H 387 (1978).
26. G. Laurence, B.A. Joyce, C.T. Foxon, A.P. Janssen, G.S. Samuel, and J.A. Venables, Surface Science 68, 190 (1977).
27. M. Naganuma and K. Takahashi, Appl. Phys. Lett. 27, 342 (1975).
28. A.Y. Cho, C.N. Dunn, R.L. Kuvas, and W.E. Schroeder, Appl. Phys. Lett. 25, 224 (1974).
29. A.Y. Cho and F.K. Reinhardt, J. Appl. Phys. 45, 1812 (1974).
30. D.P. Kennedy and R.R. O'Brien, IBM J. Res.Dev. 13, 212 (1969).

31. S.M. Sze, Physics of Semiconductor Devices (Wiley, New York, 1969) p. 115.
32. A. Fischer, K. Graf, M. Hafendörfer, H. Künzel and K. Ploog, to be published.

THE AUTHORS

Dr. Gottfried H. Döhler   Dr. Klaus Ploog

Gottfried H. Döhler graduated in Physics after his studies at the Technical University of Karlsruhe, at the Institut National des Sciences Appliquées at Lyon and the Technical University of Munich from 1958-1966. He obtained his PhD in theoretical physics with a dissertation on high field transport theory of semiconductors in 1968. Still at the Technical University of Munich, from 1969 to 1970 as an Assistant Lecturer, he was first working on the recombination kinetics of luminescence. G.H. Döhler joined the Max-Planck-Institut für Physik and Astrophysik in Munich in 1970 and thereafter the Max-Planck-Institut für Festkörperforschung in Stuttgart in 1971. His fields of main interest since have been the theory of the electronic properties of superlattices and of amorphous semiconductors. After a stay at the IBM Th.J.Watson Research Center in Yorktown Heights as an IBM World Trade Corporation Visiting Scientist from 1973 to 1975 he became a permanent staff member at the Max-Planck-Institut für Festkörperforschung.

Klaus Ploog was educated at the University of Kiel (1961-1963) and at the University of München (1964-1969) where he was graduated in Chemistry and he obtained his Ph.D. in Inorganic Solid State Chemistery, having researched the chemical vapor deposition and the structural properties of elemental boron and of nonmetal borides. From 1967 to 1969 he held the position of an Assistant Lecturer and from 1970 to 1971 he was Lecturer at the Department of Inorganic Chemistry of the University of München. During tenure of a DFG Research Fellowship at the University of Bonn he installed a four circle neutron diffractometer at the KFA Jülich and he investigated the problem of hydrogen bonding in crystalline hydrates by neutron diffraction. In 1974 he joined the Max-Planck-Institut für Festkörperforschung in Stuttgart where he has established a small group working on Molecular Beam Epitaxy of III-V compound semiconductors. The group has been particularly interested in the basic technology of the MBE process and in the MBE growth and properties of GaAs crystals with special doping structures.

SUBJECT INDEX

Amphoteric impurities 105
Auger electron spectroscopy (AES)
    5, 28, 29, 116-112, 132, 155

Carrier concentration control 33
Cells effusion 55
    isothermal double 56
    thermally linked 54
Closed tube transport 50
Congruent evaporation 17
Crucible
    open 51
    PBN 52
Crystal perfection 58
    for device performance 49
    IV-VI epitaxial layers 55
Crystal quality
    criteria by photo-diode 63

Diodes
    low-high-low GaAs IMPATT 161
    n^+-p-p^+ GaAs 160
    varactor 161
Dopants 40, 54, 127, 161
    promising n-type in MBE 159

Epitaxial IV-VI devices 50
Epitaxial growth techniques 50
Etching channels or mesas 99
Evaporative techniques 51

Furnace tantalum 52

Growth
    GaAs and related alloys 96
    interruption by nucleation 6

Helium cryopumped vacuum systems 117
Heteroepitaxy
    lattice-matched 57, 82
Heterojunctions 1, 8
Heterostructure
    D.H. lasers (double) 50, 79, 85
    IV-VI laser 79
    laser devices 136
    lattice matched 49
    multi-layer optically-pumped laser 105
    S.H. lasers (single) 79
    waveguide 98
High energy electron diffraction (HEED)
    5, 29, 90, 116
Hot wall technique 50
    for IV-VI semiconductors 81, 83

In-situ analytical equipment 15, 17, 90
    diagnostic techniques 115
Integrated optical devices (IOD) 95
    growth technique problems 109
    monolithic 96, 107
    switches, modulators, detectors 108
    tapered coupler by MBE 108
Interdiffusion coefficient 6
Ion gauge 28

Knudsen cells 24
    -type ovens 4

Subject Index

Lasers
  cw 103
  cw single mode 95
  diode 42, 50, 79
  discrete 103
  distributed feed-back 50, 88
  distributed Bragg reflection (DFB) 107
  fabrication by MBE 75
  heterostructure devices 136
    IV-VI 79
    SH (single) 50, 79
    DH (double) 1, 50, 79, 85, 103, 107
    110
  homostructure 50, 87
  injection 78
  integration 105
  significant mode discrimination 88
  single mode injection 105
  twin guide 108
Lattice constants match 55, 60
  variation of energy gap 57
Liquid phase epitaxy (LPE) 1, 87, 97
Load-lock for MBE 22, 24, 28

Molecular Beam Epitaxy (MBE) 1, 97
  as collimator 24
    volume 15, 24
    temperature 27
  crucible
    materials 15, 26
  design MBE systems 15
  epitaxy control 27
    instrumentation 17
  film uniformity 24
  flux uniformity 15
    angular distribution 25, 27
    control molecular beam 27
  furnaces 17, 23
    baffles 17, 23
    contamination 27
    materials 15
    multiple 24
    shutters 23
    thermocouples 26
  heat shielding 15
  load lock 22, 24, 26
  nipi crystals growth 155
    apparatus 156
  optoelectronic devices 49
    mask openings 104
    techniques 51
  purity of source material 20
  rectangular potential cells 105
  semiconductor surface studies 115
  shutters 15
  source baffling 15, 26
  substrate holders 15, 17, 21
    in-situ cleaning 17, 28

vacuum system 15, 17, 19

Nipi crystals 145
  apparatus 156
  electronic properties 146
  growth process 155
  high-power modulation 152
  transport and optical properties 149
Nonstoichiometry 33
  phase equilibria for binary semi-
    conductors 35
  stoichiometric deviation 40

Optical waveguides 50, 75, 77, 97
  3-dimensional 99
  heterostructure 98
  single and multistripe mesa 100
Opto-electronic devices 49, 57, 65, 70
  72, 74, 89
Overlayers 138
Oxidation of surfaces (MBE) 132

Photodiodes
  cross-striped geometry for
    Pb barrier 67
  lateral collection 74
  low noise 49
  recent developments 72
  spectral detectivity with T 68
  thin film IV-VI 65, 70
Photoemission measurements 136
Photo-voltaic i.r. detectors 42

Quadrupole mass analyser 116
Quadrupole mass spectrometer (QMS) 27

Reflection High Energy Diffraction
  (REED) 21, 24, 29, 156, 160, 164,
  165
Retrograde solubility 35

Scanning electron micrograph (SEM) 162
Schottky barrier 136, 161
Secondary ion spectroscopy (SIMS) 28,
  29, 156
Semiconductor compounds
  II-VI 17, 23, 33, 37, 39, 124, 127
  IIA-VIA 34
  III-IV 127
  III-V 2, 17, 23, 33, 37, 39, 40, 79
  IIIA-VA 34
  IV 17, 127

IV-VI 2, 17, 23, 33, 37, 41, 42, 49, 50, 51, 65, 70, 79, 81, 83, 122
 IVA-VIA 34
Semiconductor periodic superstructure potential 145
Semiconductor superlattices by MBE 3, 4, 145
 energy cap control 9
 potential 3, 8
Solidus boundaries 39
Source ovens 5, 6
Sticking coefficient 5, 19, 20, 104, 116, 128, 131, 132, 134, 156
Subbands 3, 7, 8, 9, 148, 151, 152, 153
 resonant transport 7
Substrate preparation 117
Substrates IV-VI 58
 choice p-n junctions 65
 Hall mobilities
  n-type PbTe on BaF_2 63
  PbS on SrF_2 62
 Pb chalcogenides 61
Surface analysis instrumentation 28

Surface structure 119
 adsorption and desorption 123, 135
 chemisorption 134
 comparison with composition range for GaAs 123
 HEED and LEED 116, 119, 122
 photo-emission spectra 126
 surface and interface states 136
 surface stoichiometry 120
 thermal cleaning 119

Three-terminal tunneling
 device structure 8
 ultra-high vacuum (UHV)
  pumping systems 16, 19, 21

Vacuum deposition system 52
Vacuum pumping technologies 116
 UHV 16, 19, 21
Vapour phase epitaxy (VPE) 1, 97

X-ray scattering measurements 6

COMPOUND INDEX

A 133
Ag 138
Al 100, 123, 124, 126, 127, 138, 139, 140
AlAs 96, 99, 126
(AlGa)As 1
$Al_xGa_{1-x}As$ 130
$Al_{0.17}Ga_{0.7}As$ 129
$Al_{0.17}Ga_{0.83}As$ 129
  $n-Al_{0.12}Ga_{0.88}As$ 107
  $n-Al_{0.3}Ga_{0.7}As$ 107
  $p-Al_{0.3}Ga_{0.7}As$ 107
  $p-Al_{0.12}Ga_{0.88}As$ 107
As 24, 108, 122, 123, 124, 133, 134, 137, 146, 155, 160, 162, 164, 165

BaF_2 43, 49, 60, 61, 65, 70, 71, 79
Be 105, 127, 129, 159, 160, 162
Bi 34, 41, 83, 87
Bi_2Te_3 41
BN (as pyrolytic BN) 26

C 26
CaF_2 61
Cd 104
CdS 1
$CdSe_{0.7}Te_{0.3}$ 1
CdTe 17, 39
CO 26

Ga 22, 23, 24, 27
GaAlAs 109, 124, 126, 132, 134
$Ga_{1-x}Al_xAs$ 1, 6, 16, 96, 97, 99, 103, 104, 145, 155
$Ga_{0.62}Al_{0.38}As$ 99

$Ga_{0.8}Al_{0.2}As$ 99, 105
$Ga_{0.7}Al_{0.3}As$ 102
$Ga_{0.5}Al_{0.5}As$ 105
GaAs 1, 5, 6, 24, 33, 34, 37, 39, 96, 97, 99, 100, 103, 104, 105, 108, 109, 115, 117, 118, 119, 120, 123, 124, 127, 129, 131, 132, 133, 134, 137, 139, 140, 146, 155, 156, 157, 158, 160, 164, 165
  n-GaAs 129
  p-GaAs 107
GaAsP 131
GaP 37, 39
GaSb 138, 139
Ge 40, 105, 127, 155, 157, 158, 159, 160, 162, 165

In 21
InAs 138, 139
$In_{1-x}Ga_xAs$ 6
$In_{1-x}Ga_xAs-GaSb_{1-y}As_y$ superlattice 3, 6, 8
InP 138
InP|GaInAs|InP 110

KBr 61
KCl 61
KI 61

Mg 104, 128
MgF_2 79, 88
Mn 105, 127, 159

N 27, 131

Compound Index

NaBr 61
NaCl 58, 61, 65
NaI 61

O 105, 131, 132, 133, 134

Pb 34, 43, 80
PbS 58, 61, 62, 65
PbSSe 83, 84, 85
Pb(SSe) 69, 81
$PbS_{1-x}Se_x$ 82
PbSnSe 57
(Pb,Sn)Se 70
$Pb_{1-x}Sn_xSe$ 70
PbSnTe 49, 81, 83, 84, 85, 87, 88
(PbSn)Te 65, 88
$Pb_{1-x}Sn_xTe$ 51, 70, 75, 79, 81, 88
$Pb_{0.8}Sn_{0.2}Te$ 59
$Pb_{0.92}Sn_{0.08}Te$ 75
$Pb_{0.88}Sn_{0.12}Te$ 81
$Pb_{0.78}Sn_{0.22}Te$ 87
   n-$Pb_{0.88}Sn_{0.12}Te$ 79, 83, 86
   n-$Pb_{0.78}Sn_{0.22}Te$ 79, 83
   p-$Pb_{0.88}Sn_{0.12}Te$ 81
PbTe 17, 34, 37, 41, 43, 49, 58, 59, 61, 65, 74, 75, 77, 79, 81

p-PbTe 65
PbTeSe 57

S 131, 157
Sb 127, 139
Se 123, 131
Si 1, 23, 24, 40, 100, 105, 117, 119, 120, 127, 155, 157, 158, 162
SiO_2 75, 81, 105
Sn 40, 43, 105, 127, 131
$Sn_{1-x}Pb_xSe$ 1
$Sn_xPb_{1-x}Te$ 1
SnTe 37
SrF_2 49, 61, 62

Ta 21, 25, 52
Te 34, 41, 83, 86, 87, 104, 131

Zn 104, 105, 123, 131, 159
ZnS 50
ZnSe 1, 119, 123, 136
ZnTe 1, 33, 34, 37, 39